Routledge studies in ecological economics

1 **Sustainability Networks**
Cognitive tools for expert collaboration in social-ecological systems
Janne Hukkinen

2 **Drivers of Environmental Change in Uplands**
Aletta Bonn, Tim Allot, Klaus Hubaceck and Jon Stewart

3 **Resilience, Reciprocity and Ecological Economics**
Northwest coast sustainability
Ronald L. Trosper

4 **Environment and Employment**
A reconciliation
Philip Lawn

5 **Philosophical Basics of Ecology and Economy**
Malte Faber and Reiner Manstetten

6 **Carbon Responsibility and Embodied Emissions**
Theory and measurement
João F.D. Rodrigues, Alexandra P.S. Marques and Tiago M.D. Domingos

7 **Environmental Social Accounting Matrices**
Theory and applications
Pablo Martínez de Anguita and John E. Wagner

8 **Greening the Economy**
Integrating economics and ecology to make effective change
Bob Williams

9 **Sustainable Development**
Capabilities, needs, and well-being
Edited by Felix Rauschmayer, Ines Omann and Johannes Frühmann

10 **The Planet in 2050**
The Lund Discourse of the future
Edited by Jill Jäger and Sarah Cornell

11 **From Bioeconomics to Degrowth**
Georgescu-Roegen's 'new economics' in eight essays
Edited by Mauro Bonaiuti

12 **Socioeconomic and Environmental Impacts on Agriculture in the New Europe**
Post-Communist transition and accession to the European Union
S. Serban Scrieciu

13 **Waste and Recycling**
Theory and empirics
Takayoshi Shinkuma and Shusuke Managi

14 **Global Ecology and Unequal Exchange**
Fetishism in a zero-sum world
Alf Hornborg

15 **The Metabolic Pattern of Societies**
Where economists fall short
Mario Giampietro, Kozo Mayumi and Alevgül H. Sorman

16 **Energy Security for the EU in the 21st Century**
Markets, geopolitics and corridors
Edited by José María Marín-Quemada, Javier García-Verdugo and Gonzalo Escribano

17 **Hybrid Economic-Environmental Accounts**
Edited by Valeria Costantini, Massimiliano Mazzanti and Anna Montini

18 **Ecology and Power**
Struggles over land and material resources in the past, present and future
Edited by Alf Hornborg, Brett Clark and Kenneth Hermele

19 **Economic Theory and Sustainable Development**
What can we preserve for future generations?
Vincent Martinet

20 **Paving the Road to Sustainable Transport**
Governance and innovation in low-carbon vehicles
Edited by Måns Nilsson, Karl Hillman, Annika Rickne and Thomas Magnusson

21 **Creating a Sustainable Economy**
An institutional and evolutionary approach to environmental policy
Edited by Gerardo Marletto

22 **The Economics of Climate Change and the Change of Climate in Economics**
Kevin Maréchal

23 **Environmental Finance and Development**
Sanja Tišma, Ana Maria Boromisa and Ana Pavičić Kaselj

24 **Beyond Reductionism**
A passion for interdisciplinarity
Edited by Katharine Farrell, Tommaso Luzzati and Sybille van den Hove

25 **The Business Case for Sustainable Finance**
Edited by Iveta Cherneva

26 **The Economic Value of Landscapes**
Edited by C. Martijn van der Heide and Wim Heijman

27 **Post-Kyoto Climate Governance**
Confronting the politics of scale, ideology and knowledge
Asim Zia

28 **Climate Economics**
The state of the art
Frank Ackerman and Elizabeth A. Stanton

29 **Good Governance, Scale and Power**
A case study of North Sea fisheries
Liza Griffin

30 Waste Management in Spatial Environments
Edited by Alessio D'Amato, Massimiliano Mazzanti and Anna Montini

31 The Green Fiscal Mechanism and Reform for Low Carbon Development
East Asia and Europe
Edited by Akihisa Mori, Soocheol Lee, Kazuhiro Ueta, Stefan Speck and Paul Ekins

32 Peak Oil, Climate Change, and the Limits to China's Economic Growth
Manqi Li

33 The Sustainable Economics of Elinor Ostrom
Commons, contestation and craft
Derek Wall

34 Green Industrial Policy in Emerging Countries
Edited by Anna Pegels

35 The Great Transition
Mauro Bonaiuti

36 Redefining Prosperity
Isabelle Cassiers

37 Environmental Challenges and Governance
Diverse perspectives from Asia
Edited by Sacchidananda Mukherjee and Debashis Chakraborty

38 Green Growth and Low Carbon Development in East Asia
Edited by Fumikuza Yoshida and Akihisa Mori

39 Biodiversity in the Green Economy
Edited by Alexandros Gasparatos and Katherine J. Willis

40 The Economics of Green Growth
New indicators for sustainable societies
Edited by Shunsuke Managi

41 Green Jobs for Sustainable Development
Ana-Maria Boromisa, Sanja Tišma and Anastasya Raditya Ležaić

Green Jobs for Sustainable Development

Ana-Maria Boromisa, Sanja Tišma and Anastasya Raditya Ležaić

Routledge
Taylor & Francis Group
LONDON AND NEW YORK

First published 2016
by Routledge
2 Park Square, Milton Park, Abingdon, Oxon OX14 4RN

by Routledge
605 Third Avenue, New York, NY 10017

Routledge is an imprint of the Taylor & Francis Group, an informa business

British Library Cataloguing in Publication Data
A catalogue record for this book is available from the British Library

Library of Congress Cataloging in Publication Data
Boromisa, Ana-Maria.
Green jobs for sustainable development / Ana-Maria Boromisa, Sanja Tišma and Anastasya Raditya Ležaić.
pages cm
1. Environmental sciences–Vocational guidance. 2. Green movement–Vocational guidance. 3. Sustainable development. I. Tišma, Sanja. II. Ležaić, Anastasya Raditya. III. Title.
GE60.B67 2015
363.70023–dc23 2015001811

ISBN: 978-0-415-74774-5 (hbk)
ISBN: 978-1-315-79687-1 (ebk)

Typeset in Times New Roman
by Wearset Ltd, Boldon, Tyne and Wear

For Jelena, Ivana, Juraj, Tamara and Frida.

Contents

Tables

Boxes

Abbreviations

3GF	Global Green Growth Forum
ACES	American Clean Energy and Security Act
ACTU	Australian Council of Trade Unions
ANZSCO	Australia New Zealand Standard Codes for Occupations
ANZSIC	Australia New Zealand Standard Industry Codes
ARRA	American Recovery and Reinvestment Act
ASEAN	Association of Southeast Asian Nations
AUD	Australian dollar
BIS	Department of Business, Innovation and Skills (UK)
BLS	Bureau of Labour Statistics (US)
CCAP	Climate Change Action Plan
CCL	Climate change levy
CCS	Carbon capture and storage
CEDEFOP	European Centre for Development of Vocational Training
CEFC	Clean Energy Finance Corporation
CIELAP	Canadian Institute for Environmental Law and Policy
CO_2	Carbon dioxide
DECC	Department of Energy and Climate Change (UK)
DEFRA	Department of Environment, Food and Rural Affairs (UK)
DKK	Danish krone
DWP	Department of Work and Pensions (UK)
EAF	Electric arc furnace
EC	European Commission
EEA	European Environment Agency
ELV	End-of-Life Vehicle
EGS	Environmental goods and services
EMR	Electricity market reforms
ETS	Emission Trading System
EU	European Union
FJF	Future Jobs Fund
FSDS	Federal sustainable development strategy (Canada)
G20	Group of Twenty
GBP	British pound sterling

GDP	Gross domestic product
GEF	Global Environment Facility
GESS	Green Economy Scoping Study
GGEI	Global Green Economy Index
GGGI	Global Green Growth Institute
GGND	Global Green New Deal
GGS	Green goods and services
GHG	Greenhouse gas
GIB	Green Investment Bank
H&S	Health and safety
HRSDC	Department of Human Resources and Skills Development Canada
IDDRI	Institute for Sustainable Development and International Relations (France)
ILAC	Latin American and Caribbean Initiative
ILO	International Labour Organization
IMF	International Monetary Fund
IOE	International Organisation of Employers
IPPC	Integrated Pollution Prevention and Control
ISP	Integrated steel plant
ITUC	International Trade Union Confederation
JPoI	Johannesburg Plan of Implementation
JPY	Japanese yen
KECO	Korean Employment Classification of Occupation
KOTRA	Korea Trade Centre
KSIC	Korea Standard Industrial Classification
kWh	Kilowatt-hour
LEDS	Low Emission Development Strategy
LNG	Liquefied natural gas
MDG	Millennium Development Goal
MSSD	Mediterranean Strategy for Sustainable Development
MXN	Mexican peso
NAICS	North American Industry Classification System
NDP	National Development Plan
NOC	National Occupational Classification (Canada)
NOx	Mono-nitrogen oxides
NRE	New and renewable energy
NRP	National reform programme
NSDS	National sustainable development strategy
OECD	Organization for Economic Cooperation and Development
PAGE	Partnership for Action on a Green Economy
PCSD	Presidential Commission on Sustainable Development
PV	Photovoltaic
R&D	Research and development
RAP	Regional action plan

RDA	Regional Development Agency (UK)
ROK	Republic of Korea
SDG	Sustainable development goal
SEEA	System of Environmental and Economic Accounting
SME	Small and medium sized enterprises
SNA	System of National Accounts
SNG	Synthetic natural gas
tce	tons of coal equivalent
TUC	Trade Union Congress (UK)
UK	United Kingdom
UN	United Nations
UNCTAD	United Nations Conference on Trade and Development
UNDESA	United Nations Department of Economic and Social Affairs
UNEP	United Nations Environmental Programme
UNESCAP	United Nations Economic and Social Commission for Asia and the Pacific
UNFCCC	United Nations Framework Convention on Climate Change
UNIDO	United Nations Industrial Development Organization
UNITAR	United Nations Institute for Training and Research
US	United States
USDA	US Department of Agriculture
VET	Vocational education and training
VOC	Volatile organic compounds
WEEE	Waste of Electrical and Electronic Equipment

Introduction

Green jobs, green growth and green economy have become popular in the early twenty-first century. They were used in the context of sustainable development and poverty eradication. The green economy was one of the specific themes of the United Nations Conference on Sustainable Development held in Rio in 2012 (the so called Rio+20 Conference), thus it has received significant international attention.

The declarations that followed from international institutions, scientific studies, papers and books of renowned authors indicated that a way out of the global economic crisis through preserving the environment can be found in green growth. However, implementation is far from complete.

From the global process of greening national economies, the third industrial revolution, the 'green revolution', was expected, which would save the planet and bring well-being to all people.

Green jobs have been considered as a solution for new employment, the way out of environmental, economic and social crisis, and as a specific answer to the jobless growth. The green economy was expected to support the fight against climate change, environmental degradation and poverty; and green growth was expected to reduce the gap between developed and developing countries and to ensure sustainable development.

Still, there is a vivid debate regarding whether green jobs are drivers of or obstacles to economic growth and whether they are too expensive to create. This debate is reflected in popular culture,[1] but tends to be general in nature and not sufficiently supported by numbers. The main objective of this book is to add some numbers to the debate and to outline the answers to the questions regarding to what extent the expectations of greening are achievable and feasible, and/or to what extent is greening a new marketing niche, creating a new market for new products and new consumption.

The book builds on the results of our previous research on environmental finance and development.[2] It shows, first, that significant investments are needed to protect air, water and nature (biodiversity), and to deal with waste and climate change; and second, that those investments cannot be financed solely on commercial terms, because the implementation of a polluter/user pays principle would, in some countries and sectors, exceed the level of affordability. Thus, investments in environmental services could potentially lead to increasing risk of poverty, and not necessarily to the improvement of well-being and social equity.

In this book we present wider view – in terms of geography, but also regarding role of environmental goods and services (EGS) in economic growth. We analyse to what extent the creation of new green jobs can contribute to the achievement of the post-2015 sustainable development goals (SDGs) and the implementation of a new, green growth model.

The starting hypothesis is that creating green jobs (by state investment and subsidies, through good governance and with strong political support) is not sufficient to create green growth and sustainable development.

Existing analyses show that green growth and green economy provide jobs that are better and more sustainable than ordinary jobs. The analysis provided in this book is from different a perspective: it examines to what extent the creation of green jobs supports overall economic development (as opposed to the creation of elitist jobs and a new marketing niche that may widen the gap between rich and poor).

In order to do so, general conceptual frameworks for green jobs, green economy, green growth and green policy are presented, as well as their implementation in nine countries selected from among the G20 (Group of Twenty).[3]

The focus is on the G20 members because of their size and impact, and also because the G20's policies have been criticised by anti-globalists. Going green is, to a certain extent, anti-global.

The selection includes both advanced (the European Union, the United States of America, Australia, Canada, ROK and Japan) and developing countries (Mexico, China, Turkey and Brazil). Also, the top three carbon dioxide (CO_2) emitting countries in 2012 are included, which accounted for 55 per cent of total global CO_2 emissions (China with a 29 per cent share, the United States with a 16 per cent share and the European Union with an 11 per cent share). Among the EU, the focus is on the United Kingdom and Denmark. The United Kingdom, along with other countries in the selection, is a G20 member, while Denmark is selected because of its role in promoting green policies and its image of the 'world's greenest economy': Denmark in 1971 set up the world's first Ministry of the Environment, and issued some of the world's first stringent environment regulation. It is among the world's most energy-efficient economies and has succeeded in decoupling economic growth from energy consumption.

Through these examples we try to identify the drivers for greening an economy and the prevailing concepts.

For explaining green jobs, green economy, green growth and green policies the same approach is used throughout the book. First, the general concept is described, as defined by the relevant international organisations. This is followed by a presentation of specific national approaches and concrete implementations of green policies (if any).

The analysis was prepared based on background documents and secondary sources, with the intention of developing an overall picture and framework for a global comparison. Based on common features and main differences, we identify the main driving forces of and obstacles to implementation.

The first chapter presents the concept and definitions of green jobs, as used in the international and national context. Special attention is given to discussing the number of green jobs, as well as the limitations of the estimates.

Chapters 2 and 3 deal with the green economy and green growth. As the green growth concept originated in Asia, the presentation of national approaches in Chapter 3 starts with discussing Asian countries. This chapter also includes estimates on the necessary investment to create green jobs that serve as a basis for outlining the green growth potential.

Defining sustainable development, Chapter 4 gives insights on the development of the concept and its treatment in agreements and documents such as: new post-2015 SDGs; the comparison with the Millennium Development Goals (MDGs) and their real achievements to date; the document 'The Future We Want' from the Rio+20 Earth Summit; the previous Copenhagen Climate Conference outputs, etc. It serves as a basis for the identification of the relationship between sustainable and green policies.

Green policies are the focus of Chapter 5; it shows which policies and measures are designed to achieve goals related to green jobs, green growth, green economy and sustainable development. It also describes how some well-known regular policies can be transformed into green policies, and how some of them can be used as they are for green practices. The design of indicators for monitoring the success of green policy measures is also described, with a clear purpose of highlighting important information for decision makers.

Chapter 6, on the relationship between green jobs, a green economy and green growth and sustainable development, answers the question as to whether the investment to create green jobs is wise or whether it will just create 'elite' jobs that are expensive – and may be difficult to afford at both a national and employer level. Thus, this chapter provide answers to the question of whether green growth is just a new marketing niche that may contribute to widening the gap between rich and poor, and whether the green concept is really the only way for a common sustainable future and well-being worldwide.

The final chapter presents the main findings of the book.

The analysis presented in this book shows that although the green concept is well-intentioned, its implementation depends on local circumstances – economic, political and social. Developed countries perceive green growth as a way to create new markets and demand, while developing countries rely more on labour intensive growth and less expensive green jobs. Thus, greening the economy does not diminish the differences between rich and poor.

Notes

1 For example, there is reference to green jobs/green investments in *Breaking Bad* and *House of Cards*. In *Breaking Bad*, which received an Emmy award in 2014, Skyler White alleged that she quit her accountancy job at Ted Beneke's factory because the factory was not green enough and that she returned to it when it introduced 'green welding'. In *House of Cards*, congressman Peter Russo failed to testify against the

closure of a shipyard, which led to 12,000 job losses. He tried to reduce the damage by proposing green projects for the area.

2 We (at least two of the volume's authors) could not resist this shameless self-promotion. The research results were published under the same name (*Environmental Finance and Development*) by Routledge in 2012.

3 The G20 is an international forum for the governments and central bank governors from 20 major economies: 19 individual countries – Argentina, Australia, Brazil, Canada, China, France, Germany, India, Indonesia, Italy, Japan, Mexico, Russia, Saudi Arabia, South Africa, South Korea, Turkey, the United Kingdom and the United States – along with the European Union (EU). The EU is represented by the European Commission (EC) and by the European Central Bank.

1 Green jobs

The concept

The global economic crisis has led to millions of job losses (36 million since 2007, ILO, 2014). In addition, around one-third of total employment (839 million workers) earn less than US$2/day, remaining in working poverty. Unemployment stands at more than 200 million people. Thus, at least 200 million new jobs are needed over the next five years to keep pace with the growing working-age population in emerging and developing countries (ILO, 2014a).

At the same time, there is a growing evidence on climate change. The climate change challenge has led to renewed interest in sustainable growth as a means to promote human development.

Green jobs are seen as an answer to these twin challenges. Still, despite the fact that 'green job' is an increasingly popular term, there is no single definition of a green job (e.g. Bowen, 2012; Huggins, 2009). Green jobs are very heterogeneous in terms of job skill requirements, pay levels and working conditions (OECD, 2012).

The definitions of green jobs that were put forth by various organisations (international agencies, advocacy organisations, individual state agencies) served as a basis for analysis and/or data collection. Specific analyses use available data at national or company level, and adopt terminology accordingly (e.g. low carbon jobs, clean energy jobs, environmentally dependant jobs, jobs dependant on resource efficiency improvements, green-collar jobs).

The United Nations' International Labour Organization (ILO) launched the Green Jobs Initiative in 2007 in partnership with the United Nations Environment Programme (UNEP) and the International Trade Union Confederation (ITUC); the International Organisation of Employers (IOE) joined in 2008. The aim of the Green Jobs Initiative was to assess, analyse and promote the creation of decent jobs that contribute to the implementation of environmental policies to address global environmental challenges.

For the ILO, the concept of green jobs summarises the transformation of economies, businesses, work environments and job markets towards a sustainable economy that provides decent jobs with low carbon consumption (ILO, 2014a). Relevant features of the green job relate to environmental protection, economic development and social inclusion. Thus, a green job is any job that:

1 produces a product or service related to environmental conservation and management, or
2 makes processes more sustainable within any industry, and
3 has decent work conditions (ILO, 2013c).

Hence, the green job concept combines job content and the characteristics of goods or services.

The ILO distinguishes between environmental and green jobs. Environmental jobs are those that diminish environmental impact due to their product or services, while green jobs are those that have an environmental component and include decent work conditions. Environmental jobs are not necessarily green (e.g. some jobs in recycling industries are environmental, but are often part of an unofficial economy with dangerous work conditions).

The ILO proposed a Decent Work Index in 2008 in order to analyse work conditions and screen environmental jobs and provide an estimate on the number of green jobs. The index ranges from 0 to 100. A result of 70 or higher indicates that the job can be considered decent. The index is based on conventional activities and uses four variables:

1 Adequate wage (monthly income), 50 per cent weighted.
2 Decent working hours (weekly working hours).
3 Work stability and security (written contract).
4 Social protection (access to social security).

The ILO prepared decent work country profiles for 16 selected countries,[1] while the ILO's work on green skills includes 21 countries.[2] Brazil and South Africa are included in both efforts, thus the possibilities for comparison between countries and sectors based on ILO findings is quite limited.[3]

The relevance of the 'decent' aspects of green jobs is further complicated considering the definitions of the ITUC and the United Nations Environmental Programme (UNEP), which were partners in the Green Jobs Initiative. The ITUC considers the greening of jobs (and the economy) as a means to create more and better jobs, i.e. as a way to provide decent jobs. Environmentally friendly investments should create decent jobs and help the transformation of existing jobs into sustainable jobs. The ITUC calls these jobs (sustainable and decent) green and decent jobs, implying that green jobs are not necessarily decent. The UNEP defines a green job as:

> work in agricultural, manufacturing, research and development (R&D), administrative, and service activities that contribute substantially to preserving or restoring environmental quality. Specifically, but not exclusively, this includes jobs that help to protect ecosystems and biodiversity; reduce energy, materials, and water consumption through high efficiency strategies; de-carbonize the economy; and minimize or altogether avoid the generation of all forms of waste and pollution.
>
> (UNEP, 2008)

Thus, the UNEP definition does not include a decent component.

The necessity to develop standards to enhance the international comparability of the statistics and to contribute to the measurement of decent work was addressed at the nineteenth International Conference of Labour Statisticians in 2013 (ILO, 2013c), when guidelines concerning a statistical definition of employment in the environmental sector were endorsed (ILO, 2013b: 3).

The key point of discussion before the adoption of guidelines was a link between employment in the environmental sector and decent work. It was recognised that the notion of 'green' with its emphasis on sustainability intrinsically implies that both decent work and the environmental dimensions of green jobs are indispensable for achieving sustainable development and that they should in no way be disconnected. On the other hand, there were views that introducing the dimension of decent work further complicated an already complex field of statistics. Also, it was recognised that applying a decent work approach can bring about inconsistencies with other 'green' measures that do not explicitly consider decent work, such as the System of Environmental and Economic Accounting (SEEA).[4]

Finally, according to the adopted guidelines, the term 'green jobs' refers to a subset of employment in the environmental sector that meets the requirements of decent work.

The environmental sector consists of all economic units that carry out environmental activities. These activities are broadly defined as environmental protection activities and resource management activities. The primary purpose of environmental protection activities is the prevention, reduction and elimination of pollution and other forms of degradation of the environment. The primary purpose of resource management activities is the preservation and maintenance of the stock of natural resources and, hence, safeguarding against depletion. Activities in agriculture, fisheries and forestry can be considered as environmental, if environmentally sustainable technologies and practices are used.

The requirements of decent work are adequate wages, safe conditions, workers' rights, social dialogue and social protection (ILO, 2013a: 3).

In brief, the ILO considers that green jobs are central to sustainable development and are a response to the global challenges of environmental protection, economic development and social inclusion. It promotes the greening of enterprises, workplace practices and the labour market as it considers that there is considerable potential to create decent work, associated with the transition to a low carbon sustainable development. On the other hand, the ILO identifies that transition to a low carbon economy requires restructuring and will inevitably lead to a decline in certain industries and occupations. Thus, policies should be in place to support 'just' transition and help negatively affected communities and workers (ILO, 2014b, ILO 2014c). Green jobs and related employment in the non-environmental sector created thanks to greening, employment in low carbon economic units and energy efficient enterprises, and green work could help to minimise and manage any negative effects of transition.

Employment in the non-environmental sector created thanks to greening refers to employment in economic units that supply goods and services to the environmental sector. Employment in low carbon economic units and energy efficient enterprises refers to employment in units that have low carbon emissions (for example, employment in green buildings) and to employment in enterprises that are more energy efficient than most of the enterprises within the same economic activity. Green work refers to all work involved in the production of EGS, and includes employment, voluntary work and own-use production work to produce EGS.

The key challenges raised in the discussion on the draft guidelines concerning a statistical definition of employment in the environmental sector also highlight the major differences in the definitions developed by various organisations.

Most statistical definitions take an industry approach, identifying green jobs with employment in industries that are considered to produce green products and services. They utilise a 'list approach', focussing on a subset of industries producing environmentally desirable outputs, using an OECD/Eurostat definition of EGS industry.[5] EGS industry includes the production of goods and services to measure, prevent, limit, minimise or correct environmental damage to water, air and soil, as well as problems related to waste, noise and eco-systems, covering pollution management and resource management. Differing lists of industries classified as green lead to disparate estimates of the number of green jobs.

From a different analytical perspective, green jobs are a result of green policies/green entrepreneurship. According to such an approach, the employment impacts of green policies are examined relative to a baseline. This approach allows jobs to be considered green if they are created by green policies, even if they are in sectors with no direct relationship to environmental objectives (e.g. tobacco processing), with a secondary relationship (e.g. construction). The impact of policy change should also consider job losses in polluting sectors (e.g. traditional, fossil fuelled energy).

Thus, a growing number of advocacy organisations and national governments develop their own definitions of green jobs to serve as a basis for advocacy, collecting statistics and/or making policy choices. In order to illustrate the variety of definitions across the world, in the definitions developed by the European Union and its member states, the United States (including the official definition and a selection of advocacy organisations), Japan, Australia and Korea, as well as (a lack of) effort to do so in Canada, Mexico, Brazil, Turkey and China are presented below.

The European Union and its member states

Light or dark green jobs?

In April 2012 the EU launched a set of policy documents (the so called employment package) looking into how EU employment policies intersect with a number of other policy areas in support of smart, sustainable and inclusive growth (which is the core of the EU development strategy for the period until 2020). Smart, sustainable and inclusive growth is expected to foster fundamental transformation of business processes and related skill sets, whilst also giving rise to so-called green jobs. Green jobs cover all jobs that depend on the environment or are created, substituted or redefined in the transition process towards a greener economy (EC DG Employment, 2012). Focus is on two wide categories of jobs:

1 Development of 'green jobs' in eco-industries, and
2 Supporting the 'greening' of traditional industries.

Several studies (some of them commissioned by the EC) have provided different definitions. The World Wide Fund for Nature analysed green market potential and related environmental and low carbon jobs in Europe (Ghani-Eneland *et al.*, 2009). For the purpose of the Ghani-Eneland study, green jobs were defined as jobs in the field of renewable energy, sustainable mobility, energy efficiency products and services. Ecorys[6] in its study on a number of jobs dependent on the environment and resource efficiency improvements uses wider definition, including the whole labour force in agriculture (Rademaekers, *et al.*, 2012).

On a national level, there are EU member states that have developed national definitions (e.g. Austria, Finland, Germany, Hungary, Portugal and Spain), as well as those that have not (e.g. Belgium, Croatia, Greece, France, Netherlands, Slovakia, Slovenia, Sweden). An illustrative list of existing definitions is provided in Box 1.1

Box 1.1 Illustrative list of green jobs definitions in various EU member states

The *Austrian* definition is based on the handbook on the Environmental Goods and Service Sector (EGSS) from EUROSTAT (2009).

While there is no national definition of green jobs in *Belgium*, a policy oriented approach is followed. Definitions are developed for different governments in various studies (e.g. High Employment Council, IDEA/Ecorys for the Flemish Government, IDEA for the federal government).

In 2009, *Finland* adopted a definition of green activities entirely derived from the standard classification of economic activity.

The first definition of green jobs in *Germany* was adopted in 1996. The definition has been refined several times since then. Green jobs refer to two overlapping categories: i) the production of EGS (pollution management; cleaner production and technologies; resource management – cf. Eurostat/OECD definition, 1999); ii) are involved in environment related activities (e.g. environment consultants, planning and administration, nature and landscape conservation).

In *Spain*, a qualitative definition was adopted by the Ministry of the Environment in the 1990s. The approach consists in identifying green sectors, as defined by international organisations such as the ILO, OECD or UNEP, within the national industry classification and the occupation classification.

Source: authors' compilation based on OECD (2012)

The UK's Department of Work and Pensions' (DWP) Future Jobs Fund (FJF)[7] aims to create up to 10,000 green jobs. For the purposes of this fund, a green job is defined as one that provides goods or services that help the economy to move toward lower carbon emissions and greater resource efficiency. This includes jobs in environmental sectors (including recycling, waste management, environmental consultancy and monitoring), renewable energy technologies (including wind, wave, geothermal and biomass) and emerging low carbon sectors (such as alternative fuels, carbon capture and storage (CCS), carbon finance and building technologies) (House of Commons, 2010).

The *United Kingdom* differentiates between dark green skills and light green skills, implying that there are differences between dark green jobs and light green jobs.

Generic or light green skills enable resource efficiency. These are necessary for all businesses and include: business management skills such as lifecycle analysis/costing; carbon literacy for procurement; planning, impact assessment and risk management; leadership and management; sustainable procurement; resource efficiency skills; financial management, etc.

Dark green skills support low carbon industry, climate resilience and management of natural assets. The examples of dark green jobs are scientists and engineers with training or transferable knowledge for nuclear and renewable energy (including wind and marine), technicians with training or transferable knowledge to install energy efficiency measures and retrofit at a household and business premises levels (HM Government, 2011). It is claimed that in the future every job will be a green job (CEDEFOP, 2010). While the current definition of green jobs is unclear the term is used to promote jobs, and there are specialised services for green job seekers.[8]

In *Denmark,* green jobs were established as a concept on the Danish policy agenda in the late 1990s. Although not clearly defined, green jobs have typically been associated with activities related to energy saving, recycling of waste, production of energy from renewable sources. A special public fund 'Green jobs in Denmark' that supported a wide range of activities (improved energy management in firms, the establishment of new environmentally friendly firms and products, nature preservation, waste handling and the production of organic food) was active from 1997 to 2001 (Madsen, 2104: 29). However, according to the OECD data, the Danish Ministry of Employment initiated work on a definition of green jobs only in 2012 (OECD, 2012).

The Danish Energy Agency presented survey data on the qualifications of employees in the green sectors compared to those in the rest of the economy. The survey showed that the main difference between jobs in green production and other firms is that 42 per cent of the employees with further education in firms with green production have technical education, compared to 32 per cent of other firms.

Companies with green production have only a slightly lower share of unskilled employees and a slightly higher share of skilled employees compared to other firms. Also, they are similar to other firms when it comes to the employment of persons with short-, medium- or long-cycle further education (including PhD). They are also similar to other firms in having increased the share of persons with a long-cycle education in the period 2005 to 2010. Thus, a sufficient number of graduates with backgrounds in natural sciences and technology is necessary to further develop the green sector.

The United States of America

Green-collar jobs

The official definition of green jobs was developed in 2010 by the Bureau of Labour Statistics (BLS). The process included the review of a wide range of studies, including surveys conducted by state workforce agencies, international organisations, and consultations with stakeholders (federal agencies, the state labour market information offices, industry groups).

To illustrate differences in approaches, we present the definitions of the advocacy organisation Green For All,[9] the Brooking Institute and the definition used by the US Conference of Mayors (See Box 1.2).

Box 1.2 Selected definitions of green jobs in the United States

The green or clean or low carbon economy is the sector of the economy that produces goods and services with an environmental benefit. Green jobs are jobs in this sector (Muro *et al.*, 2011.)

Green jobs (also called green-collar jobs) are well-paid, career track jobs that contribute directly to protecting or enhancing air, water and the planet. A green job has to be decent, i.e. provide a path out of poverty. A job that improves the environment but doesn't provide a family-supporting wage, path out of poverty, or a career ladder to move low-income workers into higher-skilled occupations is not a green-collar job. Most green-collar jobs are middle-skill jobs, requiring more education than high school, but less than a four-year degree, and are well within reach for lower-skilled and low-income workers as long as they have access to effective training programmes and appropriate support (Jones, 2008).

Green activities are related to generating electricity using renewable or nuclear fuels, agriculture jobs supplying corn or soy for transportation fuel, manufacturing jobs producing goods used in renewable power generation, equipment dealers and wholesalers specialising in renewable energy or energy-efficiency products, construction and installation of energy and pollution management systems, government administration of environmental programmes, and supporting jobs in the engineering, legal, research and consulting fields (US Conference of Mayors, 2008: 5).

Source: authors' compilation

The definitions by Green for All and the Conference of Mayors were used for lobbying purposes. The report commissioned by the Conference of Mayors used a list approach to identify green jobs in sectors with high growth potential at both national and metropolitan levels. The results were used for advocating investments in green industries at the metropolitan and local level, with a view to transforming the US economy (Global Insight, 2008).

The official BLS definition is based on the common features of various definitions of green jobs – green jobs are jobs related to preserving or restoring the environment, preventing and cleaning up pollution and greenhouse gases (GHGs), and conserving natural resources.

There are two components of the green job, consistent with the output and process approaches.

Green jobs are either:

1 Jobs in businesses that produce goods or provide services that benefit the environment or conserve natural resources (output approach).
2 Jobs in which workers' duties involve making their establishment's production processes more environmentally friendly or use fewer natural resources (process approach).

In the output approach, BLS is concerned with jobs related to producing a specific set of goods and services (i.e. green goods and services, GGS), and is not concerned with the environmental impact of the production process. GGS fall into one or more of five groups:

1 *Energy from renewable sources* – electricity, heat or fuel generated from wind, biomass, geothermal, solar, ocean, hydropower, landfill gas or municipal solid waste.
2 *Energy efficiency* – this group includes equipment, appliances, buildings, vehicles and products and services that improve the energy efficiency of buildings and the efficiency of energy storage and distribution (e.g. smart grids).
3 *Pollution reduction and removal, GHG reduction, and recycling and reuse* – these are products and services that:
   - Reduce or eliminate the creation or release of pollutants or toxic compounds, or remove pollutants or hazardous waste from the environment.
   - Reduce GHG emissions through methods other than renewable energy generation and energy efficiency. This category includes, for instance electricity generated from nuclear sources, the trading of certificates and offsets.
   - Reduce or eliminate the creation of waste materials and collect, reuse, remanufacture, recycle, or compost waste materials or waste water.
4 *Natural resources conservation* – this group includes products and services related to organic agriculture and sustainable forestry; land management; soil, water, or wildlife conservation; and storm water management.
5 *Environmental compliance, education and training, and public awareness* – these are products and services that:
   - Enforce environmental regulations.

- Provide education and training related to green technologies and practices.
- Increase public awareness of environmental issues.

The output approach is, to some extent, comparable to the definition of environmental jobs.

The process approach is relevant to any industry and covers activities and associated jobs that favourably impact the environment, although the product or service produced may not be 'green'. It identifies jobs in which workers' duties involve making their establishment's production processes more environmentally friendly or use fewer natural resources. These workers research, develop, or use technologies and practices to lessen the environmental impact of their establishment, or train the establishment's workers or contractors in these technologies and practices.

The purpose of the categories is to establish the scope of green jobs. They are not mutually exclusive and do overlap.

BLS's definition does not include provisions related to sustainability/decent jobs, but data include information on worker safety, wages, union membership and benefits. Thus, provided that criteria are defined to determine the necessary conditions to consider the work decent, BLS information would enable the identification of such a job.

Until 2013, BLS periodically published career information on green jobs (US Department of Labour, 2013a). The information available for occupations included: wages, expected job prospects, what workers do on the job, working conditions, and necessary education, training and credentials. However, in March 2013, BLS had to cut its budget, thus eliminating two programmes and all 'measuring green jobs' products, resulting in the loss of the following:

- Data on employment by industry and occupation for businesses that produce GGS;
- Data on the occupations and wages of jobs related to green technologies and practices; and
- Green career information publications.[10]

Japan

Green jobs fight climate change

Japan started the discussion about green jobs in the 1990s. The first estimates on green jobs focused on construction and environmental research. The Ministry of the Environment defines that the role of green business is to simultaneously fight climate change and boost the economy. Firms producing goods and services that measure, prevent, limit, minimise or correct environmental damage to water, air and soil, as well as problems related to waste, noise and ecosystems are considered green. This includes cleaner technologies and products and services that reduce environmental risk and minimise pollution and resource use. Green jobs are defined as jobs involved in the production of EGS, based on OECD definition. The reports on green jobs are regularly (every six months) provided by the Ministry of the Environment and include information on 12 distinguishing environmental industries and 19 companies engaged in environmental business (Japanese Ministry of the Environment, 2013).

The implementation of the New Growth Strategy, adopted in 2011, is expected to cause profound shifts and transitions in labour markets and the way people earn a living, and to bring major opportunities for the creation of green jobs.

Australia

First statistical experimental series on green jobs in 2014

The discussion on green jobs in Australia was launched in 2008 when the Australian Conservation Foundation defined green-collar jobs as jobs in six green industries: renewable energy, energy efficiency, sustainable water systems, biomaterial, green buildings, waste and recycling (ACTU, 2008).

A year later the Environment Institute of Australia and New Zealand released a study proposing a taxonomy of green jobs for Australia (Ehmcke *et al.*, 2009). It identified three factors that define a green-collar worker:

1 The skills and responsibilities of the individual,
2 The industry and nature of the organisation for which they work, and
3 Whether the job and the organisation tend towards the environmental or sustainable end of the green spectrum.

The proposal was based on the Australia New Zealand Standard Codes for Occupations (ANZSCO) and Australia New Zealand Standard Industry Codes (ANZSIC).

The ANZSCO occupational codes are broadly ranked from higher levels of skill or responsibility at the top to lower levels of skill or responsibility at the bottom. In the first three ANZSCO groups (managers, professionals and technicians, and trade workers) green-collar workers are usually defined by their specific skills or responsibilities, though they may also be defined by the nature of their organisation. In the last five groups, green-collar workers are almost always defined by the nature of the organisation they work for, leading to the Australian two-part definition of green-collar workers:

1 Managers, professionals and technicians who work in green organisations or who have green skills and responsibilities within other organisations that may not be considered green.
2 Services, clerical, sales and semi-skilled workers who work in green organisations.

Based on the definition, seven broad types of green-collar workers are identified: those involved in management, strategy, technology, policy, education, action and process. However, despite the definition, which was based on existing data, the Australian Bureau of Statistics has only in 2014 released an experimental series on environmental services jobs that covers organisations in green sectors closely (Emerson, 2014).

Canada

Still without a formal definition of green jobs

Defining green jobs is an ongoing and inconclusive debate in Canada. There are a number of definitions available from various organisations, including community based, non-profit corporations (e.g. Peel Halton Workforce Development Group), not-for-profit private business foundations (e.g. the Globe Foundation) or standardisation agencies (e.g. Environmental Careers Organization Canada, see Box 1.3). However, the government of Canada does not have a formal definition of green jobs.

Box 1.3 Selected definitions of green jobs in Canada

A green job is one that works directly with information, technologies, or materials that minimise environmental impact, and also requires specialised skills, knowledge, training, or experience related to these areas (Environmental Careers Organization Canada, 2010: 4).

Green occupations are those that facilitate the reduction of waste and pollution, improve the environment, and pay a liveable wage with benefits that can support a family and offer potential for upward mobility (Peel Halton Workforce Development Group *et al.*, 2009: 3)

An occupation that produces an output or lowers the price of a product that offers positive environmental externalities, may be considered in whole or in part as a green job (Globe Foundation, 2009).

Source: Katz, n.d.: 12.

The available definitions can be grouped in the following five categories, based on the emphasis that they put on certain features of green jobs:

1 Industrial-based approach that emphasises the economic outputs of a particular job.
2 Occupational-based approach that emphasises the activities and inputs of a particular job.
3 Worker-centred approach that emphasises the quality of a job first (i.e. adequate wages and opportunity for career advancement) and environmental consciousness second.
4 Energy-centred approach that includes only jobs found within energy generation and conservation.

5 All-encompassing environmental approach that includes all jobs that improve or protect the environment in some way (Katz, n.d.).

The Department of Human Resources and Skills Development Canada (HRSDC) launched an initiative on the green jobs taxonomy that combines both specific and standard nomenclatures in order to categorise industrial sectors and occupations that contribute to green economic activities. The two standard nomenclatures are the North American Industry Classification System (NAICS) and the Canadian National Occupational Classification (NOC). The green jobs taxonomy is organised into five broad green domains:

1 environmental protection;
2 green energy;
3 energy efficient construction;
4 green manufacturing and transportation; and
5 green services.

Each domain aggregates several green industry sectors. These industry sectors are based on a specific nomenclature that is then mapped to a standard industry nomenclature based on NAICS industry groupings. For each NAICS industry grouping, a selection of green occupations was identified and subsequently mapped to a standard occupational nomenclature, the Canadian NOC groupings. Progress thus far on a green jobs taxonomy is based on the research conducted by organisations including: Occupational Information Network, the BLS, the California Centre for Excellence, the United Nations Environmental Programme, the Globe Foundation, the ILO, HRSDC Service Canada LMI Regional Network, Statistics Canada and ECO (OECD, 2012).

Republic of Korea

Green jobs contribute to 'low carbon growth'

The definition of green jobs was adopted in 2009. These are the jobs involved in the production of goods and services that contribute substantially to 'low carbon growth' by achieving higher energy efficiency and by improving the environment. The definition is based on the existing Korea Standard Industrial Classification (KSIC) as well as the Korean Employment Classification of Occupation (KECO). In practice, the Korean government has classified 64 green industries out of a total 169 categories under the KECO and 101 green occupations among total 429 headings in the KECO. There are some overlapping intersections between green industries and green occupations, as the definition is not based on a separate set of classification criteria (OECD, 2012).

Republic of Korea has been implementing its 'Action Plans for Jobs – Conducive Green New Deal' scheme since January 2009. The scheme is part of an economic recovery package. A sum of KRW50 trillion being invested is expected to create 960.000 jobs that are environmentally friendly from 2009 to 2012. In addition, the Korean government in November 2009 published its longer-term 'Green Jobs Plan', designed to support the government's agenda to combine vigorous economic growth with environmental progress as well as to create green jobs and to nurture a talented green workforce. For example, the 'Green SMEs Training Consortium Programme' was developed in 2010 to raise green workforces in small and medium sized enterprises (SMEs) to effectively address skill problems in shifting to a greener economy. The designated model centre (Korea University of Technology and Education), which gives training in nine core areas including photovoltaic (PV), bio and other renewable energy sources, and eco-transport systems, is receiving temporary support, such as facilities and equipment expenses until 2016. In 2010 alone, KRW21.7 billion was budgeted to this programme, and the government expects that this will function as a bridge to channelling green technical skills in large companies such as Samsung Electronics to SMEs.

Mexico

Green jobs for a sustainable future

There is no official definition of green jobs in Mexico. The ILO's study on green jobs in Mexico used the standard ILO definition, according to which a green job is the one that:

1. Produces a product or service related to environmental conservation and management; or
2. Makes processes more sustainable within any industry; and
3. Has decent work conditions.

Traditional activities differ from green ones on the basis of the working conditions and environmental performance. Based on national standards and certification, the following nine key green activities were identified:

1. Sustainable agriculture or organic agriculture
2. Sustainable forestry activities
3. Renewable energy
4. Clean industry
5. Sustainable construction
6. Waste management (recycling)
7. Sustainable tourism
8. Public mass transport
9. Government.

The most challenging issue is related to the 'decent' component of the job. Namely, the Mexican labour market is characterised by a high rate (66.5 per cent) of informal employment, (subsistence farming, paid housework) and poor working conditions; almost half (46 per cent) of employees lack contracts and 40 per cent do not have benefits.[11]

Brazil

Green jobs for national development

Green jobs are a core element in Brazilian national development policy. The jobs are identified based on the list approach and include jobs in industries such as seedling production management, forest management, recycling, biofuel production activities and the like (*People's Daily Online*, 2009). Since 2009 the ILO has been providing permanent technical support to the conceptual development and practical implementation of a green jobs strategy, at federal and state level, based on the ILO–Brazil agreement on 'Programa empregos verdes', which was renewed in August 2010 (ILO, 2011). The work of the ILO is commendable, especially in the states of Bahia and Mato Grasso through the Decent Work Agenda.

Turkey

Green jobs estimates available only for public sector

There is no official definition of green jobs in Turkey. However, the issue has been taken up by several studies (e.g. BETAM, 2010) and the Turkish Statistics Institute (TurkStat). TurkStat provides information on the number of green-collar employees in the public sector (but not in the whole economy). Green-collar jobs in the public sector relate to energy, R&D, protecting air and climate, securing water for drinking and general consumption, waste water treatment, waste management, protecting and betterment of ground and surface waters, noise reduction, protecting biological diversity and landscape, and protecting against radiation.

In 2013 the ILO launched a two-year project aimed at assessing the existing green jobs potential and formulating a set of recommendations to contribute to a green jobs strategy for Turkey, with a particular focus on the creation of green jobs for young women and men. The project, EU assistance under vocational education, launched in 2012, targeted the employability of young workers in electricity generation. The programme provided at least 45 days of vocational training, or at least 15 days of apprenticeship for 1,000 young workers, who also attended workshops on green jobs, environment protection, energy efficiency, etc.

China

Green jobs – higher income and better job security

There are various studies dealing with green jobs in China (e.g. World Bank; 2012; ILO, n.d.; Libo, 2012; Pan *et al.*, 2011). Also, the current five-year development plan, 2011–2015, includes a section on green development. However, there is no official definition of a green job. According to the ILO, the green jobs strategy offers a win–win solution to the pursuit of growth, job creation, employment security and environmental protection in China. The ILO considered jobs in green industries and the potential for green jobs as result of the restructuring of brown industries in order to analyse employment patterns and skills for green jobs. The potential of job creation in selected sub-sectors is also assessed using a value chain assessment methodology, while the empirical analysis of low carbon economy and employment aims to quantify the number of jobs created and lost during China's transition to a low carbon economy (ILO, n.d.). Green jobs are also expected to be created by the green transformation of traditional sectors (e.g. further reduction of the energy intensity of the economy, Pan *et al.*, 2011; World Bank, 2012). While the Worldwatch study (Pan *et al.*, 2011) considers this transformative process as a main source of green jobs, the World Bank identifies two additional sources of the creation of green jobs: the expansion of emerging green industries and the expansion of the service sector (World Bank, 2012). Emerging industries are green if they emit low levels of pollution and GHGs (e.g. clean energy such as solar power, wind power, biomass, and hydropower). The rising share of services in GDP (gross domestic product) helps in reducing the economy's carbon intensity. In addition, specialised services support the creation of green jobs (e.g. ecosystem services, carbon asset management services, carbon trading, and contract energy management). It is considered that China has significant potential to develop green technology jobs that are more skilled and higher paying than current non-green technology jobs (World Bank, 2012, Libo, 2012).

Counting green jobs

Due to the vagueness of the concept of green jobs, it is hard to obtain exact information about the size of this specific form of employment (cf. the discussion of various definitions, in EC, 2009: 109–110). There is a lack of reliable, comprehensive and comparable data about green jobs in various countries. Also, time series for individual countries are limited and sometimes based on various definitions. There are countries that have not developed an official definition of green jobs, yet their statistical offices provide some data.

The studies have used different methodologies and various ways of measurement. Most of the literature focuses on direct employment created (e.g. production of equipment, maintenance), with more cursory treatment of indirect (in supply chain) and/or induced jobs. Indirect jobs are those created in the supply chain, while induced jobs are created by the consumption of those directly and indirectly employed.

Net green jobs creation can be accompanied by job losses in negatively affected industries, thus this impact should also be considered (cf. Schneck *et al.*, 2010). Most of the studies at global, regional and country level that have investigated the net impact on employment of environmental policy suggest that the impact is positive (ILO, 2013a).

According to an early estimate, in 2006 there were more than 2.3 million green jobs (compared to an employed global labour force of around 1.8 billion), including renewable energy, construction (providing improved energy efficiency in building), low carbon transport, and other environmental activities (UNEP, 2008).

Available estimates show that the number of green jobs has grown during the global financial crisis[12] and that their share in GDP is greater than the share in total jobs. Thus, increasing the share of green jobs creates growth.

The European Union and its member states

Creation of green jobs supported by environmental legislation

There has been considerable job creation in the EGS sector in the period between 2002 and 2011: from 3.0 to 4.2 million jobs between 2002 and 2011, including 20 per cent during the recession years (Eurostat, 2014).

In the same period resource productivity increased significantly (20 per cent in 2000–2011). The continuation of the trend would lead to a further increase of resource productivity (by 30 per cent by 2030), which could boost GDP by 1 per cent, while creating about two million jobs (EC, 2014: 4).

There are separate estimates of green jobs at the sector level (water, tourism, waste prevention and management). In the water sector it is estimated that 1 per cent increase of growth rate of the water industry in Europe can create between 10,000 and 20,000 new jobs (Ecorys, Acteon, 2014). Tourism and recreation in Natura 2000 sites are estimated to create eight million jobs (6 per cent of the total employment in the EU, Bio Intelligence Service, 2011). Implementing the legislation on waste prevention and management is expected to create more than 400,000 new jobs (Bio Intelligence Service, 2012), while the proposed review would create a further 180,000 jobs (EC DG Environment, 2014).

Energy efficiency policy implementation requires upgrading the skills of more than four million workers, while further use of renewables is expected to create jobs in the biomass and biofuels sectors, agriculture and forestry.

Table 1.1 shows the distribution of employment in the EU amongst the different environmental domains. The 'waste' domain has the highest share of labour (about 1.3–1.4 million people employed). The largest observed growth between 2000 and 2008 was in the 'recycling' and the 'renewable energy' sectors, with 78 per cent and 75 per cent respectively. Renewable energy has a relatively low absolute share of employment although it has the second highest growth rate.

However, there are significant differences in estimations in the various studies (see Table 1.2). The largest relate to the employment in renewable energy, where the 2000 employment figures differ by a factor of 3 (49,756 compared to 160,136 jobs), followed by the noise sector (4,176 jobs compared to 11,688, i.e. 2.8 times). The differences in 2008 are relatively smaller; the largest differences relate to the water supply (703,758 compared to 357,943 jobs, i.e. almost

Table 1.1 Total employment in EU environmental sectors

	Employment 2000	*Employment 2008*	*Employment 2010*	*Growth rate 2000–2008 (per year) (%)*	*Total growth 2000–2008 (%)*
Air	33,668	30,816	29,598	−1.10	−8.47
Waste water	390,138	418,324	417,002	0.88	7.22
Waste	1,111,613	1,361,160	1,339,923	2.56	22.45
Soil and groundwater	14,460	21,111	21,029	4.84	46.00
Noise	11,688	9,005	8,018	−3.21	−22.96
Biodiversity and landscape	40,123	47,746	53,025	2.20	19.00
Other	144,861	180,399	177,309	2.78	24.53
Water supply	375,981	367,943	348,481	−0.27	−2.14
Recycling	238,774	425,373	480,056	7.48	78.15
Renewable energy	160,136	280,394	568,002	7.25	75.10
Total	2,521,442	3,142,272	3,442,443	2.79	24.62

Source: Ecorys, 2012: 29.

Table 1.2 Comparison between old and new employment figures

	Ecorys and IDEA 2009 study			*New study; only updated EPE*		
	Employment 2000	*Employment 2008*	*Annualized growth 2000–2008 (%)*	*Employment 2000*	*Employment 2008*	*Annualized growth 2000–2008 (%)*
Air	22,600	19,067	−2.1	33,668	30,816	−1.10
Waste water	253,554	302,958	2.3	390,138	418,324	0.88
Waste	844,766	1,466,673	7.1	1,111,613	1,361,160	2.56
Soil and groundwater	14,882	18,412	2.7	14,460	21,111	4.84
Noise	4,176	7,565	7.7	11,688	9,005	−3.21
Biodiversity and landscape	39,667	49,196	2.7	40,123	47,746	2.20
Others	129,313	193,854	5.2	144,861	180,399	2.78
Water supply	417,763	703,758	6.7	375,981	367,943	−0.27
Recycle	229,286	512,337	10.6	238,774	425,373	7.48
Renewable energy	49,756	167,283	16.4	160,136	280,394	7.25
Total	2,005,764	3,441,102	7.0	2,521,442	3,142,272	2.79

Source: Ecorys, 2012:29.

double), renewable energy (1.6 times, 280,394 jobs compared to 167,283) and air sectors (1.6 times, 30,816 compared to 19,076).

Various documents identify various categories of green jobs: jobs in the EGS sector (according to the Eurostat definition) and jobs in eco-industries (according to the OECD classification), which include two broad categories:

- Small and innovative companies acting in the field of, for example, renewable energy, waste recycling, environmental auditing and consultancy.
- More capital intensive enterprises providing goods and services in specific areas, e.g. waste, waste water, transport.

Table 1.3 summarises green jobs estimates in the EU by sector.

In the *United Kingdom*, the current definition of green jobs is unclear (HM Government, 2011: 34). Up to date reliable information is still hard to come by (Walsh, 2013).

The Department of Energy and Climate Change (DECC) estimates that by 2015 the supply chain of the Green Deal could support up to 100,000 gross jobs (DECC, 2011: 2).

The EGS sector is growing. In 2004 the sector had a turnover of around GBP25 billion, and accounted for around 400,000 jobs in some 17,000 companies. Government projections suggested that the UK EGS market would reach GBP34 billion in 2010 and that it would grow to GBP46 billion by 2015, with employment growing by at least 100,000 over the same period (Forum for the Future, 2006).

The data for the low carbon and EGS for 2010 and 2011 indicate that their growth was underestimated: Green jobs contribute 3.2 per cent of all employment in the United Kingdom, i.e. 939,625 jobs. As regards geographical distribution, the largest number of green jobs is in London (17.4 per cent of all green jobs), while the highest share of green jobs in employment (4 per cent) is in Ireland (see Table 1.4). Since 2007, the number of green jobs has increased by 58,000, or about 7 per cent. This modest growth is considerable, mostly because it accrued during a period of negative economic growth. It was among the very few areas of the economy to experience growth during the global financial crisis.

As many as 1.3 million people in the United Kingdom could be working in green jobs by 2017. However, green skills shortages need to be addressed in the following areas (greenskills4greenjobs, 2013):

- Programme and project management
- Construction project management
- R&D
- Green construction – attracting and retaining the brightest and best talent
- New nuclear build – recruiting and training staff to secure the future of the UK's energy. The government want the nuclear industry and its approach to skills and workforce development to become an example of international best practice.

Table 1.3 Green jobs estimates in the EU by sector

Green sector	*Existing green jobs (2007, 2008, 2009, 2010, 2012)*	*Number of jobs/ EUR invested*	*Potential green jobs creation (2020)*	*Potential green jobs creation (2030)*	*Potential green jobs creation (2050)*	*Other data*
Eco-industries	2.9–3.6 million in 2008 1,245,614 in environmental management activities in 2008					
Energy efficiency in buildings	232,050 in EU insulation industry	25,900 total jobs/ €1 billion	261,400–378,000 new jobs			250,000 new jobs annually if eco-efficient refurbishment rates increase by 4 per cent
Renewable energy	1,114,210 in 2010	52,700 total jobs/ €1 billion	2.7 million *(based on 20 per cent RE target)*	4.4 million *(based on 45 per cent RE target)*	6.1 million *(based on 100 per cent RE target)*	
Efficient transport and sustainable mobility	900,000 in urban transport 900,000 in railway for freight and passengers 2.1 million in efficient transport	21,500 total jobs/ €1 billion				70 per cent more jobs per unit of investment in public transport than in building new road and bridge construction

Organic farming	220,000–260,000 producers in 2009 197,000 holdings in 2008	6,600 total jobs/ €1 billion			10–20 per cent more jobs per hectare than conventional farms
Biodiversity conservation		29,000 total jobs/ €1 billion			
Waste and recycling sector	2 million in 2008: 1,466,673 in waste management and 512,337 in recycling	21,300 total jobs/ €1 billion investment (70 per cent recycling rate)	2,400,000 jobs *(50 per cent recycling)* 2,963,000 jobs *(70 per cent recycling)*		Recycling 10,000 tonnes of waste creates between 6 and 12 more jobs than incineration and 25 more jobs than landfill deposition

Source: based on Gil *et al.*, 2013: 17.

Table 1.4 Geographical distribution of green jobs in the United Kingdom, 2010–2011

Region	*Jobs*	*Per cent of all green jobs*	*Per cent of all employment*
East of England	86,885	9.2	3.0
London	163,841	17.4	4.2
South East	119,858	12.8	2.9
North East	38,793	4.1	3.5
North West	93,909	10	3.0
South West	77,721	8.3	3.1
West Midlands	77,311	8.2	3.2
East Midlands	62,512	6.7	2.9
Yorkshire and Humberside	67,872	7.2	2.8
Scotland	77,703	8.3	3.1
Wales	41,506	4.4	3.1
Northern Ireland	31,714	3.4	4.0
United Kingdom	939,625	100	3.2

Source: Walsh, 2013: 2.

- Steel fixing and safety case authorship
- Attracting learners to embrace STEM careers by highlighting the opportunities available in the renewables sector.
- Renewables sector.

In *Denmark*, even before national studies have attempted to estimate the number of green jobs, most of the available international literature tended to put Denmark high up the list, when assessing the share of GDP and employment related to such 'green' activities.

Early estimates focused on the direct and indirect employment generated by activities related to the pollution and resource management industries (also called eco-industries). Direct employment covers employment in activities concerning the operation and maintenance of equipment or the provision of EGS, as well as employment in activities aimed at the production of environmental equipment or infrastructure to provide environmental services. Indirect employment covers employment in activities that provide intermediate inputs for the production of environmental equipment and services. Applying this definition, the total direct employment in eco-related jobs in Denmark has been estimated at 82,000 persons in 2000. About half of these jobs were in pollution management (Madsen, 2009).

The first national estimate was provided in 2006 and later, in 2009, a more comprehensive assessment of the number of jobs in the green economy was published, based on the concept of 'the EGS sector' (Eurostat, 2009). The number of full-time employees was estimated at 122,000 persons or around 9 per cent of the total employment in the private sector. Similar estimates have been presented in other reports (cf. the overview in Konsulenterne, 2011). The Danish Energy Agency (Energistyrelsen) has published a national statistical survey of the size of the green economy in Denmark (Danish Energy Agency, 2012).[13] The methodology follows

Eurostat's guidelines (2009). However, it covers only the private sector. A study on employment effects of investments of EUR 3.7 billion a year in promotion of sustainable energy sources (biomass, wind, waves and sun), energy saving in transportation and heating of homes and renovation of sewers and drinking water facilities estimated the overall direct effects at 50,000 jobs per year, of which 28,000 were directly related to the proposed initiatives in the above sectors. Some 22,000 additional jobs were created in the other sectors delivering goods and services to the main provider of the investments. This indirect effect was calculated applying the input–output tables of the Danish national accounts (Fagligt Fælles Forbund, 2009: 5). No assessments were made of the supplementary job generation from the additional income created by the projects.

In a similar exercise from the Economic Council of the Labour Movement, the macro-economic and employment effects from a green growth package of public investments were estimated. The investment package amounted to DKK10 billion (€1.3 billion). The proposed investments were in the areas of energy-saving renovations in public buildings and transportation. The overall employment effects in 2010 were estimated at 10,000 full-time jobs including direct effects, indirect effects and additional income effects. The calculations were based on a combination of input–output tables and the Danish macro-econometric model ADAM, which is also used by the Ministry of Finance. The net effect on the public budgets was calculated to be an increase in the deficit of DKK5 billion (€0.7 billion). This amounted to about half the initial costs of the investment package due to the additional public income from taxation and savings on unemployment benefits caused by the rise in employment.

Between 2005 and 2010, there was a decline in employment of 6 per cent, in companies with green production. This is, however, less than the overall decline in private employment.

Finally, a significant difference in the wage level between employees in the green sector and other employees can be seen. In the green sectors, the average wage level in 2010 was 9 per cent higher than the average for all firms.

The United States of America

Green jobs comprised 2.2 per cent of total employment in 2011, significantly smaller than potential figure

There are many studies that estimate green jobs in the United States. They have different approaches and methodologies, thus comparison is problematic.

In Tables 1.5 and 1.6, the US government's data from two sources are presented: the Department of Commerce (from 2010, indicating data for 2007) and data of the BLS (for 2010 and 2011).

There were 3.1 million green jobs in 2010 and 3.4 million green jobs in 2011 according to BLS. BLS planned to develop information on the number of jobs and trend over time; however, as the BLS had to make budget cuts it stopped collecting data on green jobs.

Table 1.5 Size of the US green economy by industry sector, 2007

Industry sector	*Employment (1,000s)*		*Share of total industry sector (%)*	
	Narrow	*Broad*	*Narrow*	*Broad*
Agriculture	4	4	0.3	0.3
Construction (green buildings and services)	224	304	3.0	4.1
Manufacturing	197	241	1.5	1.8
Services	1,396	1,833	1.5	1.9
All sectors	1,821	2,382	1.5	2.0

Source: US Department of Commerce, 2010.

Green jobs comprised 2.6 per cent of total employment in 2011, which is 0.1 per cent (or 157,746 jobs) more than in 2010. The bulk of jobs (almost three-quarters) are in the private sector, with the greatest numbers in manufacturing, construction, transportation and waste services. The public sector had 860,000 green jobs in 2010, accounting for about 4 per cent of public sector employment. Local government provided more than half of the public sector jobs, with the biggest sector being transportation and warehousing, including operators of mass transit systems.

GGS employment accounted for 2.3 per cent of private sector jobs (2,515,200 GGS jobs) and 4.2 per cent of public sector jobs (886,080 GGS jobs) in 2011 (Table 1.6).

Among private sector industries, construction had the largest employment increase (101,932 jobs), from 7.0 to 8.9 per cent, and reached 487,709 GGS jobs in 2011. Among the GGS activities performed within the sector are the construction of plants that produce energy from renewable sources and weatherising and retrofitting projects that reduce household energy consumption.

Manufacturing had the most GGS jobs (507,168), which accounted for 4.3 per cent of manufacturing employment (see Table 1.6). The examples of GGS produced by manufacturing industries include iron and steel from recycled inputs, air conditioning and refrigeration equipment meeting selected standards, hybrid cars and parts, and pollution mitigation equipment. GGS jobs are found in businesses that primarily produce goods and provide services that benefit the environment or conserve natural resources.

The trade sector had 223,079 GGS jobs in 2011 (1.1 per cent of the industry's employment). This sector's increase of 17,512 GGS jobs (8.5 per cent over the year), which relate to recyclable material merchant wholesalers and used merchandise stores.

Leisure and hospitality services had 23,696 GGS jobs in 2011, while GGS employment in leisure and hospitality grew by 3,054 (14.8 per cent over the year).

In the transportation and warehousing sector, the number of GGS jobs decreased by 3,382 to 238,755, a decline of about 1.4 per cent.

Table 1.6 GGS employment in the United States, 2010 and 2011

Industry	*GGS employment*		*GGS percentage*		*GGS employment change 2010–2011*
	2010	*2011*	*2010*	*2011*	
Total, all industries	3,243,533	3,401,279	2.5	2.6	157,746
Total private	2,342,562	2,515,200	2.2	2.3	172,638
Natural resources and mining	63,344	64,689	3.5	3.4	1,345
Utilities	69,031	71,129	12.5	12.9	2,098
Construction	385,777	487,709	7.0	8.9	101,932
Manufacturing	492,985	507,168	4.3	4.3	14,183
Trade	205,567	223,079	1.0	1.1	17,512
Transportation and warehousing	242,137	238,755	6.1	5.9	−3,382
Information	33,321	29,412	1.2	1.1	−3,909
Financial activities	462	475	0.0	0.0	13
Professional, scientific, and technical services	355,386	381,981	4.8	5.0	26,595
Management of companies and enterprises	62,630	69,310	3.4	3.6	6,680
Administrative and waste services	330,650	335,417	4.5	4.3	4,767
Education and health services	28,789	26,123	0.2	0.1	−2,666
Leisure and hospitality	20,642	23,696	0.2	0.2	3,054
Other services, except public administration	51,841	56,257	1.2	1.3	4,416
Federal government	208,744	213,340	7.0	7.5	4,596
State government	256,224	248,539	5.6	5.5	−7,685
Local government	436,002	424,201	3.1	3.1	−11,801

Source: BLS, 2012.

Private utilities had 71,129 GGS jobs in 2011 (12.9 per cent of total private utilities employment), mainly in electric power generation (including nuclear, hydropower, wind, biomass, geothermal and solar).

The public sector had 886,080 GGS jobs in 2011 (4.2 per cent of employment). Over the year, government GGS employment decreased by 14,890 (−1.7 per cent). Local government had 424,201 GGS jobs in 2011, the most in the public sector, representing 3.1 per cent of local government employment. The transportation and warehousing sector had the largest GGS employment in local government with 209,063 jobs.

In 2011, state government had 248,539 GGS jobs, or 5.5 per cent of state government employment. Public administration led GGS employment in state government with 164,952 GGS jobs. The enforcement of environmental regulations and the administration of environmental programmes are examples of GGS services within public administration.

The federal government had 213,340 GGS jobs, representing 7.5 per cent of federal government employment in 2011. The public administration sector had 139,884 GGS jobs, the most within federal government. The professional, scientific, and technical services sector followed with 39,714 GGS jobs.

At state level, 10 states had over 100,000 GGS jobs in 2011: California (360,245, 2.5 per cent of employment in the state), New York (266,308), Texas (227,532), Pennsylvania (167,397), Ohio (137,143), Illinois (136,447), Florida (117,433), North Carolina (108,094), Virginia (107,773), and Washington (101,593).

The District of Columbia had the highest proportion of GGS employment, at 5.1 per cent; Oregon had the second highest proportion, at 4.3 per cent.

In 2011, the majority (56.5 per cent) of GGS employment was found in establishments that exclusively produced GGS. These establishments had 1,923,251 GGS jobs. Establishments that produced a mix of green and non-GGS had 1,478,029 GGS jobs in the same time period. Establishments that produced no green goods or services (i.e. no revenue or employment associated with GGS in those establishments) accounted for 67.1 per cent of employment within the GGS scope.

The BLS data track fairly closely a Brooking Institute study (Muro *et al.* 2011), which estimated that 2.7 million Americans were employed in the clean economy. Compared to other sectors, clean industries employ more workers than the fossil fuels or bioscience industries, but fewer than information technology.

The number of green jobs is significantly smaller than the potential figure, which is estimated at between three and four million green jobs a year, out of which between 288,000 and 390,000 are in energy, between 1.3 and 1.8 million in construction, between 1.2 and 1.6 million in transport and between 251,000 and 340,000 in the manufacturing sector (Millennium Institute, 2012a). The quantification is based on the scenario that 2 per cent of GDP is investment in four key industries per country, with a 1–5-year time frame scenario. For each industry and country, the numbers of jobs have been estimated as the key indicators to analyse the impact of green economy policies and investments (Millennium Institute, 2012b).

Japan

2.2 million green jobs in 2011 predicted to double by 2020

It was estimated that in the period 1994–2000 Japan had the potential to create 18,000 jobs in construction and 17,000 jobs in environmental research. The potential is estimated on the basis of the information from Japan's Environmental Protection Agency, which assumed an annual growth rate of 8 per cent and an investment of US$959 million in these sectors (Nahmo, 2010).

In 2003, the Ministry of the Environment estimated that employment in environmental businesses in the year 2000 was 769,000, equivalent to 1.2 per cent of total employment. In 2006, employment in private environment-related businesses had already reached 1.4 million, compared to about 76,500 employed in public environmental administration (Japanese Ministry of the Environment, 2013).

In 2009, environmentally friendly businesses included renewable energy firms and makers of energy-efficient household electronics, employed some 1.4 million people and was worth about JPY70 trillion (US$745.8 billion) (Kubota, 2009, see Table 1.7).

Since 2010, the Ministry of the Environment has released its estimate of Japan's market size of environmental industry every six months. Data (based on the survey) showed that the market size had increased to approximately JPY82 trillion in 2011 (2.3 per cent up from 2010). The size of employment grew from 2.25 million people in 2010 to 2.27 million in 2011.

According to the New Growth Strategy, the expansion of environmentally friendly business is planned, involving the hiring of 2.2 million workers and boosting its value to JPY100 trillion (US$1 trillion) by 2020 (Kubota, 2009).

This expansion would require a reallocation of labour resources across sectors, and a careful assessment of the net employment effects, while the labour market and education policies should take into account the new skills and competences that will be needed to adjust to greener technologies, greener production processes and greener working methods. The Ministry of the Environment indicated a possible reduction in employment in some environmental sectors in the 2010–2011 period (e.g. waste water management, renewable energy).[14]

Table 1.7 Environmental jobs in Japan, 2000–2011

	Number of jobs (millions)	*Value, trillion yen*
2000	0.8	30
2009	1.4	70
2010	2.25	80
2011	2.27	82
2020 (plan)	4.4	100

Source: authors' compilation based on Sumikura, 2009; Capozza, 2011.

Australia

20 per cent of green jobs are in renewable energy

According to the Australian Conservation Foundation, six industries that are considered green (e.g. renewable energy, energy efficiency, sustainable water systems, biomaterial, green buildings, waste and recycling) employed 112,000 green-collar workers in 2008 (ACTU, 2008). Almost 20 per cent of which (about 21,000 people) are employed in the renewable energy sector. Although using very broad categories of 'Environmental and Agricultural Scientists' and 'Environmental and Occupational Health and Safety (H&S) Professionals', the Department of Education, Employment and Workplace Relations reported (in 2008) that these categories had some 300,000 employees in 2007, had experienced a growth of some 3,500 over the previous five years, and that future job prospects were expected to be good (Australian Department of Education, Employment and Workplace Relations, 2008).

The 2008 Growing the Green-collar Economy report by the Australian government's Commonwealth Scientific and Industrial Research Organisation found that policy action to support the transition to an environmentally sustainable society could increase employment in sectors with high potential environmental impacts by 230,000–340,000 new jobs over the next 10 years. For the period until 2030 the potential is estimated at 770,000 extra jobs, primarily in clean energy, public transport and sustainable land management policies. Another view on the estimate of green job numbers comes from predictions of job possibilities. The Australian government's activities related to green jobs primarily focus on skills development, rather than green job creation. Australia's Green Skills Agreement, a partnership between the federal government, states and territories, was established in 2009 with the aim of strengthening the capacity and capability of the national vocational education and training (VET) sector. Under the Green Skills Agreement, all new trade apprentices commencing training after 1 January 2010 graduate with a core set of green skills, while 30,000 apprentices in carbon exposed industries graduating by late 2011 have qualifications that include clean and green skills, implying that practically all new trade apprentices require a core set of green skills and that at least 30,000 jobs in carbon exposed industries should become green. In addition, between January 2010 and December 2011 job subsidies (AUD79.6 million) were provided for 26 weeks to employers in environment rehabilitation and protection activities and emerging green and climate change industries, hiring 10,000 unemployed people aged 17–24 (OECD, 2012).

Canada

Green jobs comprise 4 per cent of total employment – mostly younger people working in waste management and science

Although an official definition of the green economy has not been established, Environmental Careers Canada (ECO Canada) has developed a working definition using national employment data and labour market trends.

According to their data, green employment has seen a tenfold increase over the past two decades. There were more than 730,000 green jobs in Canada in 2013, representing just over 4 per cent of the total Canadian labour force.[15]

Compared with 2010, the number of environmental professionals has grown by about 7 per cent (ECO Canada, 2013). Almost 90 per cent of all environmental employees work in the following 5 industries:

1 Administrative and support, waste management and remediation services (25.1 per cent).
2 Professional, scientific, and technical services (22.4 per cent).
3 Wholesale trade and retail trade (13.9 per cent).
4 Other services (13.8 per cent), and
5 Manufacturing (13.7 per cent).

It is estimated that about 460,000 Canadian establishments (19.6 per cent of all Canadian establishments) employ at least one environmental employee (up from 17 per cent in 2010).

On average, the environmental workforce is slightly younger than the overall Canadian workforce, is well-educated and has high levels of experience and responsibility. Approximately, two-thirds (67.0 per cent) of Canadian environmental employees are between the ages of 30 and 54. Over a third (36.6 per cent) hold at least a Bachelor's degree and a managerial role (37.1 per cent), while only one-quarter (25.2 per cent) of environmental employees occupy entry-level or junior-level positions (ECO Canada, 2013).

Republic of Korea

70 per cent of people with green jobs have advanced skill levels

There are several estimates on the number of green jobs in Korea for the period since 2006 (see Table 1.8). Data for 2006 relate to jobs in environmental industries: In 2006, the total number of employees in environmental industries was 390,406, of which 178,174 employees (45.6 per cent) were directly related to environment fields (Lee *et al.*, 2010, see Table 1.8).

Table 1.8 Employment and green employment in Korea, 2006–2014

	2006	*2007*	*2009*	*2012*	*2013*	*2014*
Total jobs	n/a	23,749,643	24,000,000	n/a	n/a	n/a
Green jobs	178,174	285,382	610,000	502,382	480,000	810,000
Per cent	n/a	1.21	2.5	n/a	n/a	n/a
Green related jobs	212,222	1,066,360	n/a	n/a	1,420,000	n/a
Per cent	n/a	4.49	n/a	n/a	n/a	n/a

Source: authors' compilation based on Lee *et al.* (2010) and Chang *et al.* (2011).

The Ministry of Labour estimated the number of green jobs in 2007 on the basis of the National Standard of Industry and National Standard of Occupation. From a total of 169 industries (with 3-digit codes), 47 industries and 53 occupations out of 139 total occupations (with 4-digit codes) were identified as being related to the green economy. In addition, if those industries indirectly related to green sectors are included, then 73 industries and 114 occupations could fall into the green category (Lee *et al.* 2010, see Table 1.9).

It was estimated that over 70 per cent of people with green jobs have advanced skills. Among green-related jobs, there are 42 per cent with a mid-level range of skills and 40.4 per cent with advanced skills. As a result, green workers are likely to be better paid.

The Ministry of Labour has also projected the development of green jobs by industry for the period 2007–2013 (Table 1.10).

It has been projected that number of green and green-related jobs in agriculture will decline, while in the manufacturing and services sectors growth is projected. Overall, a net positive effect is expected to prevail (2,275 jobs lost, while 179,417 gained).

The Korean government's classification used for its 2009 estimate is based on standard industrial classification. The classification identifies 164 industrial subsectors, which include green jobs. There are 101 occupational categories of 'green jobs', which are aggregated in five wide groups (Chang *et al.*, 2011) (Table 1.11).

The total number of green jobs in 2009 was estimated to 610,000, or 2.5 per cent of employment.[16] Of this, the 'energy sources' sector provided 24,000 green jobs; the 'enhancing energy efficiency' sector 54,000; the 'greening industry and space' sector 251,000; the 'protecting the environment and recycling resources' sector 102,000; and the 'low carbon economic activities' sector 179,000 (see Table 1.11).

However, the same study provided a different estimate for the development of green jobs: It was estimated that by 2012 about 8,000 social jobs in environment fields would be created and 35,000 jobs in the field of wind and solar energy.

By 2012, 100,000 new jobs in renewable energy, with another 100,000 in energy efficiency and 17,000 in waste-to-resources sectors were created, and

Table 1.9 Green jobs and green-related jobs by sector in Korea, 2007

	Total 2007	*Agriculture, fishing, mining*	*Manufacturing*	*Services*
Total jobs	23,749,643	1,913,515	4,144,775	17,691,353
Green jobs	285,382	10,578	114,157	161,647
Per cent	1.21	0.55	2.75	1.09
Green related jobs	1,066,360	23,948	519,338	523,074
Per cent	4.49	1.25	12.53	2.96

Source: Lee *et al.*, 2010.

Table 1.10 Green jobs prospects by industry

		2007	*2008*	*2009*	*2013*	*Increase/ decrease*	*Annual average increase/decrease*	*Annual average increase rate*
Green jobs	Agriculture, fishing, mining	10,578	12,690	9,855	10,415	−2,275	−455	−3.4
	Manufacturing	114,157	117,293	121,150	145,835	28,542	5,078	4.5
	Services	169,756	183,965	206,961	325,840	141,875	28,375	12.1
	Total	294,492	313,948	337,976	482,091	168,142	33,628	9.0
Green related jobs	Agriculture, fishing, mining	23,948	26,434	24,580	26,216	−217	−43	−0.1
	Manufacturing	519,338	537,774	536,781	606,737	68,963	13,793	2.5
	Services	523,074	551,248	576,236	790,778	239,530	47,906	7.5
	Total	1,066,360	1,115,456	1,137,597	1,423,732	308,276	61,655	5.0

Source: Lee *et al.*, 2010: 51.

Table 1.11 Composition of green employment in Korea, 2009

Sector	*Number of jobs*
Energy sources	24,000
Enhancing energy efficiency	54,000
Greening industry and space	251,000
Protecting the environment and recycling resources	102,000
Low carbon economic activities	179,000
Total	610,000

Source: authors' compilation based on Chang *et al.*, 2011.

Table 1.12 Comparison of number and growth rates in various studies, Korea

	Growth rate (2009–2013)	*Number of new jobs created*	*Jobs in 2014*
Green jobs (Chang *et al.*, 2011)	6	200,000	810,000
Green jobs (Lee *et al.*, 2010)	9	136,000	480,000
Green-related jobs (Lee *et al.*, 2010)	5	n/a	1,420,000

Source: author's compilation from Chang *et al.*, 2011 and Lee *et al.*, 2010.

520,000 green jobs in total (see Table 1.12). The renewables industry alone is expected to generate 950,000 new jobs by 2030.

The studies estimate various annual growth rates – between 5 and 9 per cent, depending on the type of job (see Table 1.12) and the study itself. The total number of estimated jobs differs significantly, as different definitions of green and green-related jobs are used. According to Chang *et al.*, the average rate of the increase of the number of green jobs is estimated at 6.0 per cent from 2009 to 2013, with the number of green jobs expected to reach 810,000 in 2013. That is, 200,000 new green jobs are expected to be created in four years.

It was estimated that in period 2009–2013 the number of green jobs will increase by 9.0 per cent annually (34,000 jobs), reaching 480,000 green jobs in 2014. Green related jobs were estimated to increase by 5 per cent (to an estimated total of 1,420,000 jobs in 2013) (Lee *et al.*, 2010).

Thus, it can be concluded that the statistics on green jobs in Korea would benefit from some improvement in order to enable trend identification and comparison of the results of various studies.

Mexico

Green jobs from renewable energy to sustainable tourism

There is a total of 1.815 million direct environmental jobs in Mexico, which represent about 4.5 per cent of the workforce in 2011 (ILO, 2013a).

There is a lack of quantitative employment data related to activities designed to reduce carbon emissions and pollution, promote energy and resource efficiency and prevent the loss of biodiversity and ecosystem services.

Among nine key green activities, the sectors that show a greater percentage of environmental jobs out of the total workforce in each sector are: renewable energy (22 per cent), recycling (12 per cent), sustainable construction (11 per cent), sustainable forestry and reforestation (8 per cent) and clean industry (7 per cent). The sectors that have a penetration of environmental employment of 6 per cent or less are organic agriculture (6 per cent), sustainable tourism (2 per cent) and government (2 per cent).

Brazil

Huge potential for greening jobs in agriculture, fisheries, nature protection and tourism

Brazil had more than 2.6 million green jobs in 2008, accounting for 6.73 per cent of the total formal jobs in the country. The jobs considered as green in the ILO's study included activities such as seedling production management, forest management, recycling and biofuel production activities.

The huge number of people working as recyclable materials collectors was not included, because of their lack of social protection and health hazards they are exposed to.

There is a great potential for greening jobs in sectors based on exploiting natural resources and advancing environmental quality, such as mining, marketing, building maintenance and use, agriculture, livestock and fisheries, as well as tourism, all of which formally employ about 5.8 million people.

Turkey

Less than 1 per cent green jobs in public employment

Figures from the Turkish Statistics Institute (TurkStat) show that the number of green-collar employees in the public sector was 7,557 in 2008. Only 67 per cent of those employees worked exclusively in jobs related to environmental protection, while the remaining had other duties as well. The number of green-collar employees increased in the period from 1997 until 2005, when it peaked at 14,500 and accounted for a 0.4 per cent share in public employment (0.06 per cent in total employment). The decrease in numbers since then has been attributed partly to the closure of the General Directorate of Rural Services and the transference of its duties to the provincial administrations.

In 2015, the results of the ILO's two-year project, including a national green jobs assessment is due to be published and will be used for the development of a set of recommendations to contribute to a Green Jobs Strategy for Turkey.

China

New emerging green sectors: forest parks management and tourism

The three leading greening sectors of China's economy are: energy, transportation and forestry, which also enable greening jobs.

Increasing the share of renewable energy in the energy sector (primarily solar thermal energy, PV and wind power) can significantly contribute to the creation of green jobs. In the period 2006–2010, China's solar PV power sector generated annually about 9,200 jobs (2,700 direct jobs and 6,500 indirect jobs) (see Table 1.13).

This is projected to increase to an average of 23,050 jobs annually in the period 2011–2020 (6,680 direct jobs and 16,370 indirect jobs).

China's wind power industry – both the power generation and turbine manufacturing sectors – created an average of 40,000 direct green jobs annually between 2006 and 2010. China's wind power development between 2011 and 2020 is projected to generate some 34,000 green jobs annually on average. Workers at wind firms have higher average annual incomes and better job security, have experienced better occupational conditions and enjoyed a higher level of workplace protection measures than their counterparts in conventional power plants (ILS and Chinese Ministry of Human Resources and Social Security, 2010).

According to the World Bank's projections until 2030, exports of green technologies and services specifically related to renewable energy and clean energy (mainly electric) vehicles will rise to US$229–395 billion in export sales and 4.4–7.8 million new jobs (World Bank, 2012: 244).

In the transport sector, the Chinese market for alternatively fuelled vehicles is expanding rapidly. By mid-2010, there were about 5,000 such vehicles in China.

Table 1.13 Greening energy, transportation and forestry sectors in China

	2007–2010	*2011–2020*
Solar PV	9,200	23,500
Direct	2,700	6,680
Indirect	6,500	16,370
Wind power	40,000	34,000
Vehicles		1,200,000
High speed rail		630,000
Direct		230,000
Indirect		400,000
Urban rail in Beijing		437,000
Forestry	1,600,000	1,100,000
New forests		1,000,000
Forest parks		1,000,000
Total green jobs in 2020		4,500,000

Source: authors' compilation based on Pan *et al.*, 2011.

Due to the continuation of the government's policy to prioritise the development of hybrid and electric vehicles during the period 2011–2020, production could reach 1.67 million vehicles annually, which would lead to the creation of roughly 1.2 million green jobs annually.

Further development of high-speed rail would create an average of 230,000 direct and 400,000 indirect jobs annually during the period 2011–2020, or a total of 630,000 jobs annually. Urban rail development in Beijing with targets to complete 660 km of lines by 2015 (at a total investment of US$77 billion) and the construction of another 340 km of lines during the period 2016–2020 (at a total investment of US$69 billion) could bring more than 437,000 jobs each year by 2020.

It is estimated that the forestation sector employed as many as 1.8 million full-time workers in 2010 alone, or an average of 1.6 million workers annually during the 2005–2010 period. To achieve its 2020 goals, China's forestation activities could offer as many as 1.1 million direct and indirect jobs annually during 2011–2020. This employment potential should not be regarded as new jobs, due to the short term nature of forestation work. After forestation, forest management is needed to preserve and maintain the newly created green capital. Forest management jobs are long-term positions. Managing the newly added forest area during the period to 2020 would bring another million jobs.

China also has great potential to expand its niche sector of forest park tourism, as the country is home to more than 2,000 forest parks nationwide. By 2020, this relatively new green sector could provide 392,000 direct jobs and 607,000 indirect jobs, or nearly one million green jobs in total.

On the other hand, due to increasing agricultural productivity and shifting China's economy away from a labour-intensive model, rural unemployment is expected to continue rising. New job opportunities in China's emerging green sectors will help relieve the social burden of unemployment in more traditional economic sectors. Also, increased energy efficiency may lead to job losses (increasing efficiency by 60 per cent is estimated to lead to about 17.3 million job losses). However, new jobs in the low carbon and services sectors could outstrip these losses by almost ten million (Global Climate Network, 2009).

Due to the different methodologies used to derive green jobs estimates, it may not be appropriate to simply add these estimates together to derive an overall green jobs figure for China. It can be stated with confidence, however, that the three sectors – energy, transportation and forestry – could provide at least 4.5 million green jobs in 2020.

'Green' economic activities create exports and jobs but also imply some higher costs, industrial restrictions and layoffs, as well as government actions leading to changes in prices and production patterns. The positive impact on employment is greater the longer the time frame being considered and the wider the definition of 'green jobs' being used. A study under the China Council of International Co-operation on Environment and Development (Development Research Center of the State Council, World Bank, 2013: 229) estimated that government spending of RMB5.8 trillion (US$91 billion) on measures to save energy, protect the environment and replace polluting industries with hi-tech firms would create 10.6 million jobs over 5–10 years. In contrast, eliminating the dirtier sectors of the economy would lead to the loss of 950,000 jobs.

Although rough, all of these estimates confirm the notion that the net trade-offs favour greening as a source of job creation.

Final remarks

The debate on the definitions of green jobs is inconclusive. The approach adopted by international labour statisticians is that green jobs refer to a subset of employment in the environmental sector that meet the requirements of decent work (ILO, 2013b: 31).

However, additional terms are also used in an attempt to highlight social, economic or/environmental components of green jobs, such as:

- environmental and decent;
- green and decent;
- green and sustainable, which implies that green jobs are not by definition decent or sustainable.

Jobs outside environmental sectors created as result of the greening are considered as green work or light green jobs (Table 1.14).

The empirical results show that that most green-collar jobs are middle-skill jobs, requiring more education than high school, but less than a four-year degree. Companies with green production have only a slightly lower share of unskilled employees and a slightly higher share of skilled employees than other firms. Thus, such jobs are within reach for lower skilled and low-income workers, provided that they have access to training.

Table 1.14 Differentiation between green jobs and green work

Sector	*Green jobs*	*Green work*
	Environmental	*Non-environmental*
Characteristics of job	Environmental, decent, sustainable	Result of green job
Type of activities	Environmental protection activities	
Resource protection activities	Used for production of environmental goods and services	
Terms used interchangeably	Dark green jobs Green and decent jobs	Light green jobs

Source: authors' compilation.

At a national level, there are significant differences in approaches.

In some countries green jobs are counted, in some countries green work, while there are approaches that associate them with the activities in low carbon industries. Similar wording for different concepts leads to confusion and makes cross-country comparisons unreliable. Green jobs are expected to create green work, improve social well-being and to drive economic growth.

The available estimates show that number of green jobs is growing. This trend is mainly the result of the more rigorous environmental legislation in most countries, constant new international initiatives, environmental contamination remediation, management and monitoring in the sectors of waste, water, air and nature protection. However, the numbers provided are not completely coherent, with definitions of green jobs provided by international labour statisticians and relating particularly to jobs in the energy sector, but also in other sectors (e.g. tourism). The decent component of the green job definition requires continuous updating; this requires structural change and the development of the green economy.

Notes

1 These countries are: Armenia, Austria, Azerbaijan, Bangladesh, Brazil, Cambodia, Cameroon, Ethiopia, Indonesia, Jordan, Niger, Pakistan, Philippines, Senegal, South Africa and Tanzania.

2 These are: Australia, Brazil, Bangladesh, Denmark, China, Mali, Estonia, Costa Rica, Uganda, France, Egypt, Germany, India, Spain, Indonesia, United Kingdom, Philippines, United States, ROK, South Africa, Thailand.

3 Variables depend on the country and sector, e.g. there is no absolute national level of adequate wage – it depends on sector and skills. Decent working hours vary among sectors (e.g. manufacturing and services, such as tourism).

4 The SEEA enables the production of internationally comparable statistics on the environment and its relationship with the economy. The SEEA framework follows a similar accounting structure to the System of National Accounts (SNA) and uses concepts, definitions and classifications consistent with the SNA in order to facilitate the integration of environmental and economic statistics.

5 It should be noted, however, that the OECD has not endorsed a specific definition of green job (OECD, 2012: 23).

6 Ecorys is an international company providing research, consultancy and management services, specialising in economic, social and spatial development.

7 The FJF was introduced in October 2009 to support the creation of subsidised jobs for unemployed young people (primarily 18–24 years old) who were at a disadvantage in the labour market. Official statistics indicated that between October 2009 and March 2011, just over 105,000 jobs were created under the FJF. The programme cost approximately GBP680 million (i.e. about GBP6,476 per job).

8 Such as the GreenJobs, environmentjob, Renewable Energy Jobs and GreenJobsOnline websites, online, available at: www.greenjobs.co.uk/, www.environmentjob.co.uk/jobs, www.renewableenergyjobs.com/uk, and www.greenjobsonline.co.uk/.

9 Green For All was established in 2007 by Van Jones to lobby for the US' federal climate, energy, and economic policy initiatives. Van Jones is the author of The Green-collar Economy and the founder (and currently board member) of Green For All, a national organisation working to get green jobs to disadvantaged communities. He co-hosts Crossfire on CNN and is the president (and co-founder) of Rebuild the

Dream, a 'think tank' that promotes innovative solutions to fix the US economy. Van Jones was the main advocate for the Green Jobs Act, the first piece of federal legislation to codify the term 'green jobs', signed into law by George W. Bush in 2007. In 2009 he worked as the green jobs advisor to the Obama White House. More information can be found online, available at: vanjones.net.

10 Based on the Balanced Budget and Emergency Deficit Control Act, the BLS had to cut its budget by more than US$30 million, about 5 per cent of the 2013 appropriation.

11 The total employment in Mexico in 2011 reached 40.5 million jobs. Of the 40.5 million workers, 66.5 per cent (26.9 million) work informally.

12 According to the UNEP (2008) there were 2.3 million green jobs globally in 2006. In 2012, the EU's eco-industry alone was employing 3.4 million workers.

13 The difference in results is roughly 15 per cent: The estimated number of employees in the green sector is 106,000 full-time persons, of which 55,900 are in manufacturing, 46,600 in services, and 3,500 in agriculture. Total green employment equals 8.5 per cent of all employees in private firms with at least 1 employee. The largest share of green employment is found in manufacturing, where 15 per cent of employees are in firms with green production. For agriculture the share is 12 per cent, and for services 6 per cent.

14 For more details see Table 2.3 in the next chapter.

15 ECO Canada uses term environment professional for an individual who spends at least half (50 per cent) of his/her work time performing activities related to the environmental industry. Environmental employees are defined as workers who spend at least some of their time (but not necessarily half of their time) on environmental activities.

16 The number of the entire employed population is 24 million, and the unemployment rate is 3.4 per cent.

References

ACTU (Australian Council of Trade Unions), 2008. *Green Gold Rush: How Ambitious Environmental Policy can make Australia a Leader in a Global Race for Green Jobs*. Australian Conservation Foundation. Online, available at: www.acfonline.org.au/sites/default/files/resources/Green_Gold_Rush.pdf (accessed 20 November 2014).

Asia Business Council, 2009. *Addressing Asia's New Green Jobs Challenge*. Online, available at: www.asiabusinesscouncil.org/docs/GreenJobs.pdf (accessed 30 October 2014).

Australian Department of Education, Employment and Workplace Relations, 2008. *Australian Jobs 2008*. Canberra.

Bahçeşehir University Center for Economic and Social Research (BETAM), 2010. 'Turkey's EU Accession to Increase Green Jobs and Employment'. In *Today's Zaman*, 4 April. Online, available at: www.todayszaman.com/business_turkeys-eu-accession-to-increase-green-jobs-and-employment_206337.html (accessed 11 November 2014).

BIO Intelligence Service, 2011. Estimating the Economic Value of the Benefits Provided by the Tourism/Recreation and Employment Supported by Natura 2000, Final Report prepared for EC–DG Environment.

Bowen, A., 2012. '"Green" Growth, "Green" Jobs and Labor Markets'. *Policy Research Working Paper* 5990. Washington: World Bank. Online, available at: http://elibrary.worldbank.org/doi/pdf/10.1596/1813-9450-5990 (accessed 4 December 2014).

Cadogan-Cowper, A. and Johnson, T., 2011. *Measuring the Green Economy*. Australian Bureau of Statistics. Online, available at: http://unstats.un.org/unsd/envaccounting/londongroup/meeting17/LG17_23b.pdf (accessed 5 December 2014).

Capozza, I., 2011. 'Greening Growth in Japan', *OECD Environment Working Papers*, No. 28, OECD Publishing. Online, available at: www.oecd-ilibrary.org/environment/greening-growth-in-japan_5kggc0rpw55l-en (accessed 20 October 2014).

CEDEFOP (European Centre for Development of Vocational Training), 2010. *Skills for Green Jobs Country Report, United Kingdom*. Online, available at: http://ilo.org/wcmsp5/groups/public/--ed_emp/--ifp_skills/documents/publication/wcms_142471.pdf.

Chang, Y., Han, J. and Kim H., 2011. *Green Growth and Green New Deal Policies in Korea: Are they Creating Decent Green Jobs?* A Paper for the GURN/ITUC workshop on 'A Green Economy that Works for Social Progress' Brussels, Belgium, 24–25 October 2011. Online, available at: www.ituc-csi.org/IMG/pdf/Green_Growth_and_Green_New_Deal_Policies_in_Korea_ECPI_.pdf (accessed 26 October 2014).

Cruz Caruso, L.A., 2010. *Skills for Green Jobs in Brazil*. Unedited background country. ILO. Online, available at: http://ilo.org/wcmsp5/groups/public/--ed_emp/--ifp_skills/documents/publication/wcms_142300.pdf (accessed 28 October 2014).

Danish Energy Agency, 2012. *Green Production in Denmark – and its Significance for the Danish Economy*. Online, available at: www.ens.dk/sites/ens.dk/files/policy/green-production-denmark-contributes-significantly-danish-economy/Green%20production%20in%20Denmark%20-%20web%20111212.pdf (accessed 21 November 2014).

Development Research Center of the State Council, World Bank, 2013. *China 2030: Building a Modern, Harmonious, and Creative Society*. World Bank.

EC, 2011. *EU Environment Policy Supporting Jobs and Growth*. Luxembourg. Online, available at: http://ec.europa.eu/environment/enveco/industry_employment/pdf/facts_and_figures.pdf (accessed 16 November 2014).

EC DG Employment, 2012. *The Jobs Potential of a Shift towards a Low-carbon Economy*. Online, available at: www.oecd.org/els/emp/50503551.pdf (accessed 28 September 2014).

EC DG Employment, Social Affairs and Inclusion, 2013. *European Employment Observatory Review, Promoting Green Jobs throughout the Crisis: a Handbook of Best Practices in Europe*. Online, available at: www.google.hr/url?sa=t&rct=j&q=&esrc=s&source=web&cd=1&ved=0cdoqfjaa&url=http%3a%2f%2fec.europa.eu%2fsocial%2fblobservlet%3fdocid%3d10295%26langid%3den&ei=46iovly_g6nfygpnnikidg&usg=afqjcngzajw9m2yasfahurp_nyfomgvtfw&sig2=21laj-lgo6p-qfxm27h5xa&bvm=bv.81828268,d.bgq&cad=rja (accessed 10 December 2014).

EC DG Enterprise and Industry, 2014. *Study on the Competitiveness of the EU Renewable Energy Industry, Policy Analysis and Sector Summaries*. Online, available at: http://bookshop.europa.eu/en/study-on-the-competitiveness-of-the-eu-renewable-energy-industry-pbnb0214951/?catalogcategoryid=fmekabst_fqaaaej0pey4e5l (accessed 28 September 2014).

EC DG Environment, 2014. *Impact Assessment for 2014 Waste Review*. Staff Working Documents. Online, available at: http://ec.europa.eu/environment/waste/target_review.htm.

ECO Canada, 2010. *Defining the Green Economy*. Online, available at: www.eco.ca/pdf/Defining-the-Green-Economy-2010.pdf (accessed 5 November 2014).

ECO Canada, 2013. *Profile of Canadian Environmental Employment*. Online, available at: www.eco.ca/ecoreports/pdf/2013-Profile-Canadian-Environmental-Employment-ECO-Canada.pdf (accessed 10 October 2014).

Ecorys, 2009. *Study on the Competitiveness of EU Eco-industry*. Online, available at: http://ec.europa.eu/enterprise/newsroom/cf/itemdetail.cfm?item_id=3769&tpa_id=203&lang=en (accessed 5 November 2014).

Ecorys, Acteon, 2014. *Potential for Sustainable Growth in the Water Industry Sector in the EU and the Marine Sector – Input of the European Semester*. Project.

Ehmcke, W., Philipson, G. and Kold-Christensen, C., 2009. *Who are the Green-collar Workers*? Environment Institute of Australia and New Zealand. Online, available at: www.eianz.org/sb/modules/news/attachments/71/Green%20Collar%20Worker%20report%20Final.pdf (accessed 10 November 2014).

Emerson, M., 2014. 'Green Jobs Grow as GFC Recedes'. *Sidney Morning Herald*. Online, available at: www.smh.com.au/business/green-jobs-grow-as-gfc-recedes-20141101-11fepz.html (accessed 2 December 2014).

Eurostat, 2014. *European Statistics*. Online, available at: http://ec.europa.eu/eurostat.

Fagligt Fælles Forbund, 2009. *Green Jobs: Examples of Energy and Climate Initiatives that Create Employment.* Cited in Madsen, P.K., 2014. Employment in the Green Economy in Denmark –Statistics and Policies.

Feldbaum, M., 2009. *Going Green: the Vital Role of Community Colleges in Building a Sustainable Future and Green Workforce.* National Council for Workforce Education and the Academy for Educational Development. Online, available at: www.fhi360.org/sites/default/files/media/documents/Going%20Green_The-vital-role-of-community-colleges.pdf (accessed 14 October 2014).

Ghani-Eneland, M., Renner, M. and Chawla, A., 2009. *Low Carbon Jobs for Europe, Current Opportunities and Future Prospects.* WWF. Online, available at: www.google.hr/url?sa=t&rct=j&q=&esrc=s&source=web&cd=1&ved=0cb4qfjaa&url=http%3a%2f%2fassets.panda.org%2fdownloads%2flow_carbon_jobs_final.pdf&ei=jbipvntahsrrus_7glae&usg=afqjcnewcjubuzovho_hneqg4vv2vwazga&sig2=qeb2yijbg1ub9_rb2nihvq&bvm=bv.81828268,d.d24&cad=rja (accessed 2 December 2014).

Gil, B.M., Sanchez Lopez, A.B. and Murillo, L.M., 2013. *Green Jobs and Related Policy Frameworks: an Overview of the European Union.* Sustainlabour. Online, available at: www.sustainlabour.org/documentos/Green%20and%20decent%20jobs-%20An%20Overview%20from%20Europe%20FINAL.pdf (accessed 12 November 2014).

Global Climate Network, 2009. *Low Carbon Jobs in an Interconnected World.* IPPR. Online, available at: www.ippr.org/publications/low-carbon-jobs-in-an-interconnected-world.

Globe Foundation, 2010. *Careers for a Sustainable Future: a Reference Guide to Green Jobs in British Columbia.* Online, available at: http://bcgreeneconomy.globeadvisors.ca/media/4858/globe_green_jobs_guide_final.pdf (accessed 28 October 2014).

HM Government, 2011. *Skills for a Green Economy, a Report on the Evidence*. London, UK. Online, available at: www.gov.uk/government/uploads/system/uploads/attachment_data/file/32373/11-1315-skills-for-a-green-economy.pdf.

House of Commons, 2010. *Green Jobs and Skills: Government Response to the Committee's Second Report.* London. Online, available at: www.publications.parliament.uk/pa/cm200910/cmselect/cmenvaud/435/435.pdf (accessed 16 November 2014).

Huggins, L. (ed.), 2007. *Seven Myths about Green Jobs*. Online, available at: http://perc.org/sites/default/files/ps44.pdf (accessed 2 November 2014).

ILO, no date. *Green Jobs in China*. Online, available at: www.ilo.org/wcmsp5/groups/public/--asia/--ro-bangkok/--ilo-beijing/documents/publication/wcms_133679.pdf (accessed 9 November 2014).

ILO, 2011. *Green Jobs Country Brief Brazil.* Online, available at: www.ilo.org/wcmsp5/groups/public/--dgreports/--integration/documents/publication/wcms_149662.pdf (accessed 2 December 2014).

ILO, 2012. *Working towards Sustainable Development: Opportunities for Decent Work and Social Inclusion in a Green Economy*. Online, available at: www.ilo.org/wcmsp5/

groups/public/--dgreports/--dcomm/--publ/documents/publication/wcms_181836.pdf (accessed 1 November 2014).

ILO, 2013a. *Evaluation of the Potential of Green Jobs in Mexico*. Online, available at: www.ilo.org/wcmsp5/groups/public/--ed_emp/--emp_ent/documents/publication/wcms_236143.pdf (accessed 1 November 2014).

ILO, 2013b. *Report of the Conference*. 19th International Conference of Labour Statisticians Geneva, 2–11 October 2013. Online, available at: www.ilo.org/wcmsp5/groups/public/--dgreports/--stat/documents/publication/wcms_234124.pdf (accessed 10 October 2014).

ILO, 2013c. *Statistics of Work, Employment and Labour Underutilization*. Report for discussion at the 19th International Conference of Labour Statisticians (Geneva, 2–11 October 2013). Online, available at: www.ilo.org/wcmsp5/groups/public/--dgreports/--stat/documents/publication/wcms_220535.pdf (accessed 1 November 2014).

ILO, 2013d. *Sustainable Development, Decent Work and Green Jobs*. Report V. Online, available at: www.ilo.org/wcmsp5/groups/public/--ed_norm/--relconf/documents/meeting document/wcms_207370.pdf (accessed 10 October 2014).

ILS and Chinese Ministry of Human Resources and Social Security, 2010. *Study on Green Employment in China*. Beijing: ILO. Online, available at: www.ilo.org/wcmsp5/groups/public/---asia/---ro-bangkok/---ilo-beijing/documents/publication/wcms_155395.pdf.

Japanese Ministry of the Environment, 2013. *Estimated Market Size of Japan's Environmental Industries of 2011 and Report of Japan's Environmental Industry Growth Engine*. Online, available at: www.env.go.jp/en/headline/headline.php?serial=1936 (accessed 1 November 2014).

Jones, V., 2008. *Green-collar Jobs mean Standing up for People and the Planet*. Grist. Online available at: http://grist.org/article/memo-to-candidates/.

Katz, J., no date. *Emerging Green Jobs in Canada: Insights for Employment Counsellors into the Changing Labour Market and its Potential for Entry-Level Employment*. Online, available at: http://ceric.ca/files/PDFs/Emerging%20Green%20Jobs%20in%20Canada.pdf (accessed 2 December 2014).

Konsulenterne, N., 2011. *Green Employment: a Literature Survey for the Ministry of Employment*. Copenhagen.

Lee, S.Y., Jin, M.S. and Song, C.Y., 2010. *Skills for Green Jobs in the Republic of Korea*. Background country study. ILO. Online, available at: www.ilo.org/wcmsp5/groups/public/--ed_emp/--ifp_skills/documents/publication/wcms_142476.pdf (accessed 2 November 2014).

Libo, W., 2012. *Green Jobs in China: Comparative Analysis, Potentials and Prospects*. Friedrich Ebert Stiftung. Online, available at: www.fes-asia.org/media/publication/2012_Green_Jobs_Study_CHINA.pdf (accessed 1 November 2014).

Madsen, P.K., 2009. *The Employment Dimension of Economy Greening, Denmark*. EEO Review. Aalborg: European Employment Observatory.

Madsen, P.K., 2014. 'Employment in the Green Economy in Denmark –Statistics and Policies'. In Larsen, C., Rand, S., Schmid, A. and Keil, R. (eds), 2014. *Sustainable Economy and Sustainable Employment: Approaches to measuring sustainability and local labour market monitoring*. Copenhagen: Rainer Hampp Verlag.

Millennium Institute, 2012a. *Millennium Institute Methodology*. Online, available at: www.ituc-csi.org/millennium-institute-methodology.

Millennium Institute, 2012b. *United States Country Profile*. Online, available at: www.millennium-institute.org/ (accessed 3 October 2014).

Muro, M., Rothwell, J. and Devashree, S., 2011. *Sizing the Clean Economy: a National and Regional Green Jobs Assessment*. Brookings Institute. Online, available at: www.brookings.edu/research/reports/2011/07/13-clean-economy (accessed 5 October 2014).

Nahmo, G., 2010. 'Green Economics and Green Jobs: Implications for South Africa'. In Aravossis, K. and Berbbia, C.A., *Environmental Economic and Investment Assessment III*. Ashurst, UK: WIT Press: 257–269.

OECD (Organization for Economic Cooperation and Development), 2012. *The Jobs Potential of a Shift towards a Low-carbon Economy*. Online, available at: www.oecd.org/els/emp/50503551.pdf (accessed 3 December 2014).

Pan, J., Ma, H. and Zhang, Y., 2011. *Green Economy and Green Jobs in China*. Worldwatch Report. Online, available at: www.worldwatch.org/system/files/185%20Green%20China.pdf (accessed 29 October 2014).

Peel Halton Workforce Development Group, Toronto Workforce Innovation Group and Workforce Planning Board York Region, 2009. *Greening the Economy: Transitioning to New Careers*. Online, available at: http://durhamworkforceauthority.ca/wp-content/uploads/2012/10/Green_Occupations_Profiles_Final_2011.pdf (accessed 20 November 2014).

People's Daily Online, 2009. Brazil over 2.6 mln Green Jobs in 2008: ILO Report. Online, available at: http://en.people.cn/90001/90777/90852/6838181.html.

Rademaekers, K., van der Laan, J., Widerberg, O., Zaki, S., Klaassens, E., Smith, M. and Steenkamp, C. 2012. *The Number of Jobs Dependent on the Environment and Resource Efficiency Improvements*. Rotterdam: Ecorys. Online, available at: http://ec.europa.eu/environment/enveco/jobs/pdf/jobs.pdf (accessed 2 December 2014).

Schneck, J., Murray, B.C., Gumerman, E. and Tegen, S., 2010. 'Estimating the Employment Impacts of Energy and Environmental Policies and Programs'. Duke Nicholas Institute, Workshop Summary Report. Online, available at: http://nicholasinstitute.duke.edu/sites/default/files/publications/estimating-employment-impacts-paper.pdf.

Reuters, 2009. *Japan Mulls Expanding Green Business Market, Jobs*. Online, available at: www.reuters.com/article/2009/01/07/idUST327766 (accessed 20 November 2014).

UNEP, 2008. *Green Jobs: Towards Decent Work in a Sustainable, Low-carbon World*. Online, available at: www.unep.org/PDF/UNEPGreenjobs_report08.pdf.

US Conference of Mayors, 2008. *Current and Potential Green Jobs in the US Economy*. US Metro Economies. Online, available at: www.usmayors.org/pressreleases/uploads/GreenJobsReport.pdf (accessed 2 September 2014).

US Department of Labor, 2013. *Green Goods and Services News Release*. Online, available at: www.bls.gov/news.release/ggqcew.htm (accessed 21 November 2014).

Walsh, K., 2013. *Promoting Green Jobs throughout the Crisis. UK. European Employment Observatory*. Online, available at: www.eu-employment-observatory.net/resources/reviews/UK-EEO-GJH-2013.pdf (accessed 21 November 2014).

World Bank, 2012. *China 2030*. Online, available at: www.worldbank.org/content/dam/Worldbank/document/China-2030-complete.pdf (accessed 21 November 2014).

2 Green economy

The concept

The concept of a green economy is receiving increasing international attention as countries explore new patterns of development that take into account economic, social and environmental sustainability considerations.

The term green economy appeared for the first time in a 1989 report for the government of the United Kingdom entitled *Blueprint for a Green Economy* (UN, 2012). The report dealt with the definition of the term 'sustainable development' and its implications for the measurement of economic progress and the appraisal of projects and policies. Apart from the title, the report had no further reference to a green economy.

The concept of a 'green economy' re-emerged in the early twenty-first century amid the convergence of several interrelated global crises (economic, social and environmental).

In 2008, UNEP launched its Green Economy Initiative to provide analysis and policy support for investment in green sectors and for greening resources, and revived the term in the context of discussions on the policy response to multiple global crises. As part of the Green Economy Initiative, UNEP commissioned Edward Barbier, an original author of the Blueprint for a Green Economy, to prepare a report entitled a Global Green New Deal (GGND). The GGND, released in 2009, proposed a mix of policy actions that would stimulate economic recovery while improving the sustainability of the world economy. The GGND called on governments to allocate a significant share of green stimulus funding to green sectors and set out three objectives:

1 Economic recovery,
2 Poverty eradication, and
3 Reduced carbon emissions and ecosystem degradation.

It also proposed a framework for green stimulus programmes as a part of economic recovery efforts.

Despite numerous initiatives, governments failed to make any agreement on climate change at the 2009 United Nations Framework Convention on Climate

Change (UNFCCC) negotiations. It became obvious that governments were unlikely to commit to significant policy change if it was going to hinder economic growth and international competitiveness.

In 2010, ministers and heads of delegation of the UNEP Global Ministerial Environment Forum acknowledged that the green economy concept 'can significantly address current challenges and deliver economic development opportunities and multiple benefits for all nations'. Following the UN's General Assembly resolution 64/235, which defined that green economy in the context of sustainable development and poverty eradication would form one of the two specific themes for Rio+20, green economy received a great deal of international attention. Numerous reports tried to further define and demystify the concept (e.g. UNEP, 2011; the UN Environment Management Group, 2011;[1] UNEP, United Nations Conference on Trade and Development (UNCTAD), and United Nations Department of Economic and Social Affairs (UNDESA, 2011).

Rio+20's outcome document, *The Future We Want*, urges governments to 'develop their own green economy strategies through a transparent process of multi-stakeholder consultation' (§38). It recognises that green economies are not a 'rigid set of rules' (§27) and must support developing country governments to eradicate poverty (§42).

Despite significant attention and numerous attempts, there is no internationally agreed definition of green economy, and at least ten separate definitions have been identified recently (see Box 2.1).

Box 2.1 Selected definitions of green economy

1 A green economy is one that results in improved human well-being and social equity, while significantly reducing environmental risks and ecological scarcities. It is low carbon, resource efficient and socially inclusive. In a green economy, growth in income and employment should be driven by public and private investments that reduce carbon emissions and pollution, enhance energy and resource efficiency, and prevent the loss of biodiversity and ecosystem services (UNEP, 2011).
2 A system of economic activities related to the production, distribution and consumption of goods and services that result in improved human well-being over the long term, while not exposing future generations to significant environmental risks or ecological scarcities (UNEP, 2008).
3 An economy that results in improved human well-being and reduced inequalities, while not exposing future generations to significant environmental risks and ecological scarcities. It seeks to bring long-term societal benefits to short-term activities aimed at mitigating environmental risks. A green economy is an enabling component of the overarching goal of sustainable development (UNCTAD, 2011).
4 A green economy is 'a resilient economy that provides a better quality of life for all within the ecological limits of the planet' (Green Economy Coalition, 2011).

5 A green economy is described as an economy in which economic growth and environmental responsibility work together in a mutually reinforcing fashion while supporting progress on social development (International Chamber of Commerce, 2011).
6 The green economy is not a state but a process of transformation and a constant dynamic progression. The green economy does away with the systemic distortions and dysfunctionalities of the current mainstream economy and results in human well-being and equitable access to opportunity for all people, while safeguarding environmental and economic integrity in order to remain within the planet's finite carrying capacity. The economy cannot be green without being equitable (Danish 92 Group, 2012).
7 The green economy involves largely new economic activities and must provide an important entry-point for broad-based black economic empowerment, addressing the needs of women and youth entrepreneurs and offering opportunities for enterprises in the social economy (Government of South Africa, 2011).
8 The green economy can be seen as a lens for focusing on and seizing opportunities to advance economic and environmental goals simultaneously (Rio+20 Objectives and Themes of the Conference – UNCSD, 2011).
9 An economy where investment in sustainable production and cleaner technologies is shaped by key principles of social justice, social protection and decent work is considered green (ITUC, 2012).
10 The green economy is considered as one of the important sustainable solutions to solve and rectify the impacts of development on the environment, especially when our environmental resources were affected negatively by the massive economic development (UAE Ministry of the Environment and Water, 2013).

Source: authors' compilation and UN (2012)

As described above, UNEP has defined the green economy as 'one that results in improved human well-being and social equity, while significantly reducing environmental risks and ecological scarcities. It is low carbon, resource efficient, and socially inclusive' (UNEP, 2011). Another definition for green economy offered by the Green Economy Coalition (a group of non-governmental organisations, trade union groups and others) succinctly defines green economy as 'a resilient economy that provides a better quality of life for all within the ecological limits of the planet'. Some definitions include a 'list approach', i.e. they consider that the green economy is composed of a subset of industries producing environmentally desirable outputs, using the OECD/Eurostat definition of EGS industry (OECD, 1999). EGS include the production of goods and services to measure, prevent, limit minimise or correct environmental damage to water, air and soil, as well as problems related to waste, noise and the eco-system, covering pollution management, resource management.

According to available estimates, the global green markets are estimated at US$1.4 trillion per year (UNEP, 2008). Clean technology (which is only a part of the green market and includes nine broad subsectors: biofuels and biochemical, power generation, energy infrastructure, energy efficiency, industrial process efficiency and abatement, recycling and waste, remediation, transportation, water and waste water) is expected to be the third largest global industrial sector by 2020, worth US$3 trillion (Analytical Advisors, 2011)

Recognising the potential of the green economy to foster growth and development, the UN partner agencies – the United Nations Environment Programme (UNEP), the ILO, the United Nations Industrial Development Organization (UNIDO) and the United Nations Institute for Training and Research (UNITAR) – launched the Partnership for Action on a Green Economy (PAGE), with a focus on national capacity development.

PAGE was established in 2013 as a response to the outcome document of the United Nations Conference on Sustainable Development (Rio+20), entitled *The Future We Want*, which recognises the green economy as a vehicle for sustainable development and poverty eradication. PAGE aims to support 30 countries in the period until 2020 in building national green economy strategies that will generate new jobs and skills, promote clean technologies, and reduce environmental risks and poverty. During the first two years of the partnership, PAGE will focus on the following pilot countries: Burkina Faso, Ghana, Mongolia, Peru and Senegal. For each country its own green economic pathway will be forged, based on its national priorities and natural assets.

The ability of national actors to act on the green economy is key to effective policy making and achieving tangible results. On a global level, the data show that real GDP growth was faster than the increase in energy related emissions and that carbon intensity declined, showing a decoupling of economic growth from energy consumption (Table 2.1). However, to reach the emission targets necessary to limit global warming to 2°C, decarbonising at 6 per cent a year is necessary (compared to the 1.2 per cent achieved). Getting on track to limit emissions, the global economy should speed up changes in carbon intensity – it should decarbonise at 6.2 per cent a year until 2100, which is more than five times faster than now.

Thus, approaches to the green economy on regional and national level are presented below. In order to enable comparison, the global green economy index (GGEI) is used.[2]

National approaches

The EU and its member states

Green economy in Europa 2020

For the EU the 'green economy' is a model that

> secures growth and development, safeguards human health and well-being, provides decent jobs, reduces inequalities and invests in, and preserves biodiversity, including the ecosystem services it provides (natural capital), for

Table 2.1 Decarbonisation progress in 2013

Country	*2012–2013*				*Five-year trend*	
	Change in energy related emissions	*Real GDP growth (PPP)*	*Carbon intensity (tCO2/2013$m)*	*Change in carbon intensity*	*Annual average change in carbon intensity*	*Average change in GDP*
	2012–2013	*2012–2013*	*2013*	*2012–2013*	*2008–2013*	*2008–2013*
World	1.8	3.1	323	↘−1.2	−0.6	2.8
G7*	1.2	1.3	281	→−0.2	−1.9	0.6
E7**	3.5	5.4	404	↘−1.7	−0.3	6.1
Australia	−4.7	2.7	338	↓−7.2	−4.6	2.4
UK	−3.2	1.7	206	↘−4.8	−2.9	0.1
Italy	−5.9	−1.9	172	↘−4.1	−2.9	−1.6
China	3.4	7.7	561	↘−4.0	−1.6	8.9
South Africa	−1.1	1.9	635	↘−3.0	−3.0	1.8
EU	−2.5	0.1	209	↘−2.5	−2.3	−0.2
Canada	−0.6	2.0	375	↘−2.5	−1.9	1.3
South Korea	0.5	3.0	435	↘−2.4	0.2	3.0
Japan	−0.8	1.5	285	↘−2.3	−0.5	0.3
Argentina	1.7	3.0	190	↘1.2	−2.4	4.5
Saudi Arabia	2.8	3.8	380	→−1.0	0.7	4.2
Turkey	3.3	4.0	217	→−0.7	−1.4	3.7
Russia	0.6	1.3	458	→−0.7	−0.7	1.0
Mexico	0.7	1.1	231	→−0.3	0.6	1.6
Indonesia	5.7	5.8	206	→−0.1	1.3	5.9
France	0.5	0.2	145	→0.3	−2.0	0.1
US	2.5	1.9	326	→0.6	−2.4	1.0
India	6.0	5.0	271	→0.9	−0.7	6.7
Germany	3.3	0.4	224	↑2.9	−0.9	0.6
Brazil	8.1	2.5	157	↑5.5	1.8	2.6

Source: PWC, 2014.

Notes

* G7 includes Canada, United States, Japan, Germany, France, United Kingdom and Italy.

** E7 stands for seven emerging countries: China, Russia, India, Mexico, Indonesia, Brazil and Turkey.

its intrinsic value and for its essential contribution to human well-being and economic prosperity.

(European Parliament and the Council, 2013)

The European development strategy, Europa 2020, focuses on the transition towards a green, low carbon and resource efficient economy in achieving smart, sustainable and inclusive growth. The greening of the economy is mainly driven by a long-term global trend of increasing resource scarcity.

The wording of the set goals – green, low carbon and resource efficient economy – to a degree implies that the green economy is not necessarily low carbon and resource efficient.

Other EU strategic documents, such as the Resource Efficiency Roadmap, also include elements of the green economy concept. Sector specific strategies and policies further elaborate key elements, such as the decoupling of growth and resource use.

There is no comprehensive analysis of the green economy and its size in the EU. Data on eco-industry, which is among the biggest sectors in the EU, is used as proxy. Before the global financial crisis, annual turnover of eco-industries was about €319 billion and was growing by around 8 per cent per annum (Ecorys *et al.*, 2009). There are significant differences among sectors – there are mature sectors and those with rapid growth.

According to the European Environment Agency (EEA), the most relevant sectors for greening the economy relate to environmental and resource policy areas, namely:

- energy;
- GHG emissions and ozone-depleting substances;
- air quality and air pollution;
- transport sector emissions of GHGs and air pollutants;
- waste;
- water;
- sustainable consumption and production;
- chemicals;
- biodiversity and land use (EEA, 2013).[3]

The EEA has monitored Europe's progress in greening the economy since 2012. It uses using six key indicators to assess resource efficiency (related to nitrogen emissions, carbon emissions, air pollution, maritime use, water use and material resource), and a further six addressing ecosystem resilience (relating to loss of biodiversity, climate change, air quality, marine environment, water stress and material resource use). The findings indicate a mixed performance, with more progress in improving resource efficiency than preserving ecosystem resilience.

Greening the economy is seen as a way for improving productivity and competitiveness: Meeting the EU's energy goals (20 per cent of energy from renewable sources, a 20 per cent increase in energy efficiency) could save €60 billion on Europe's bill for oil and gas imports by 2020. In addition, further integration of the European energy market can boost GDP by 0.6 to 0.8 per cent.

The *United Kingdom* is committed to greening the economy and delivering long-term sustainable growth. The commitment can be illustrated by adapting the Climate Change Act, 2008, which obliges the United Kingdom to significantly reduce its carbon emissions by 2050. The Green Economy Council was set up in 2010 and the Green Investment Bank (GIB), the first bank of this type, in 2012. Since 2010, annual conferences on green skills for green jobs have been held, which focus on green skills and the role they play in strengthening the UK's low carbon economy.

According to the UK government, a green economy is not a sub-set of the economy at large – the whole economy needs to be green. A green economy is the one that maximises value and growth across the whole economy, while managing natural assets sustainably (HM Government, 2011b: 4).

The main features of the UK's green economy are:

1 sustainable and long-term growth, reduction of emissions and other environmental impacts, and the creation of green jobs (including in a low carbon and EGS sector);
2 efficient use of natural resources;
3 resilience (i.e. the United Kingdom will be less reliant on fossil fuels whilst maintaining secure supplies of energy and other natural resources);
4 exploration of comparative advantages.

The strategy for transition to a green economy presents activities of the UK government to ease the transition, which include the promotion of international action; regulation; financial incentives; voluntary agreements; fiscal measures; public sector procurement; the provision of information; and targeted work to unblock non-financial barriers to the deployment of clean energy technologies.

It is expected that the green economy will enable sustainable, long-term growth, the reduction of emissions and other environmental impacts; the creation of green jobs (including in a low carbon and EGS sector); maintaining the global competitiveness of the UK industry; the efficient use of natural resources as well as efficient demand management and efficiency measures for energy and other resources (House of Commons, 2009).

Progress towards the low carbon target is slower than expected, which can be explained by continuing economic difficulties and complicated responsibilities. There are three main government departments with green agendas: the DECC, the Department of Environment, Food and Rural Affairs (DEFRA) and the Department of Business, Innovation and Skills (BIS). In addition, there are administrations in Scotland, Wales and Northern Ireland dealing with green issues.

The EGS sector is growing. In 2004 the sector had a turnover of around GBP25 billion. In 2010 it was estimated at GBP34 billion and in 2015 GBP46 billion (House of Commons, 2009).

The green economy (which is wider than EGS) is estimated to be worth more than GBP120 billion a year (€143 billion) or 8 per cent of the UK's GDP (Greenskills4greensjobs, 2013).

Significant investment in skills and staff development is required to sustain this level and create new opportunities in low carbon businesses (e.g. manufacturing, energy, waste management, building, research and financial services). For a new, low carbon business and government infrastructure to be established, the necessary investment range is between GBP200 billion and GBP1 trillion over the period to 2030.[4] It is estimated that traditional sources of capital for investment in green infrastructure can provide less than half the necessary amount by 2025 (GBP50 to 80 billion) (House of Commons and Environmental Audit Committee, 2011).

The GIB was established to support investments in the UK's green economy. In order to meet environmental targets, the necessary investment in the energy system and other key green sectors, such as transport, waste, water and flood defences by 2020 is GBP330 billion (House of Commons and Environmental Audit Committee, 2011). Since its formal launch in November 2012, the GIB has backed 34 green infrastructure projects, committing over GBP1.4 billion and helping to mobilise GBP4.9 billion into the UK's green economy (GIB, 2014).

Transition to the green economy is supported by the Low Carbon Industrial Strategy and its Low Carbon Transition Plan. This includes the first investments from the GBP405 million for low carbon industries and advanced green manufacturing announced in the 2009 budget (House of Commons, 2011). Together with the increasing share of renewable energy, economic activity will have to move to the low carbon sectors.

Industries that produce products or services with high embedded emissions (e.g. steel making) will need to reduce the carbon footprint of their goods, while industries that produce carbon-dependent products (e.g. petrol car manufacturers) will need to diversify and change their products. A study funded by the EC found that under some economic scenarios the net effect on jobs in the UK energy sector could be negative. Therefore the Trade Union Congress (TUC) has called for greater monitoring of this displacement of jobs, because it is not clear that the government has a coherent approach to deal with this economic displacement. The TUC considers that government's approach should reflect the ways that employment in new and traditional sectors will be affected. Regional Development Agencies (RDAs) are in a position to provide some assessment of this regional and sectoral displacement.

The Low Carbon Industrial Strategy makes reference to collaboration between RDAs and other partners, but does not detail how this will happen. Trades unions have argued for a 'just transition' to a low carbon economy that fairly distributes the costs and benefits of policies across the economy.

According to the research carried out by the Environmental Change Institute at the University of Oxford into building firms, product manufacturers and suppliers could stand to tap into a new market worth between GBP3.5 billion and GBP6.5 billion per year if the United Kingdom developed policies, skills programmes and financial incentives to upgrade the existing housing stock to make it greener and more energy efficient (House of Commons, 2009: 17).

The United Kingdom gradually improves its ranking in each successive GGEI edition, but still lags behind its northern European and Nordic countries. Generally, the country ranks quite well in the perception survey (eighth), exceeding the corresponding performance values (twentieth).[5]

Denmark has decided to lead the transition to a green growth economy and to be the first country in the world to be independent of fossil fuels by 2050. The Danish government in partnership with the Confederation of Danish Industry, the Danish Energy Association, the Danish Agriculture and Food Council and the Danish Wind Industry Association established the State of Green, public–private partnership to promote green practice in ten green sectors: energy efficiency, heating and cooling, intelligent energy, wind power, solar and other renewables, bioenergy, water, climate adaptation, resources and environment, and sustainable transport. Danish best practices on greening economy relate to water policy, energy policy, and spatial planning.

The importance of green production for the Danish economy is exemplified by the fact that green production in 2010 accounted for 9.2 per cent of the total turnover in Danish firms with at least one employee. The export of 'green products' amounts to 10.4 per cent of total exports, an increase of 1 per cent since 2005. Denmark is a world leading wind power manufacturer. In 2013 Danish companies supplied 25 per cent of the world's wind turbines. The wind industry makes an important contribution to the Danish economy, employing some 27,500 people, with exports amounting to about DKK50 billion (US$9.2 billion) in 2013 (WWF, 2014).

The green economy has also been hit by the crisis. Between 2005 and 2010 there was a decline in employment of 6 per cent in companies with green production. This is, however, less than the overall decline in private employment.

Denmark continues to maintain its leading position on the GGEI, both in terms of perception and performance. A true green country brand, Denmark relentlessly communicates its commitment to green growth through a variety of strategies and tactics, and global practitioners continue to signal the positive impact of these efforts. Denmark generally ranks higher in perception compared to other Nordic nations, particularly as a vital market for clean-tech investment and innovation. Greater environmental focus on fisheries and forests coupled with translating Denmark's strong green policies into greater efficiency improvements in sectors like buildings and transport will likely improve Denmark's score further in the future.

The United States of America

Green energy as a proxy for green economy

Many studies have estimated the contribution of green businesses and jobs to the US economy. These differ in their definition of green, their methodology, and their coverage area. Such differences make comparisons across studies problematic.

Despite these differences, there are common features in many of the studies as to what constitutes green. Most studies utilise the concepts of energy efficiency, renewable or clean energy sources, and the protection of the environment as guiding principles for defining the green economy. These principles are also echoed in the environmental product lists of Eurostat (2014) and the Organization for Economic Cooperation and Development (OECD, 1999).

Although the similarities are strong, differences in opinion and definition can arise. For instance, there is disagreement on whether nuclear electricity generation should be considered a green business.

Data produced by the US Department of Commerce show that green products and services comprised 1 to 2 per cent of the total private business economy in 2007 (Table 2.2). The lower estimate was developed using a narrow definition that included products that generated little debate regarding their 'greenness'. The larger estimate was based on a broad definition that included products that some might argue were not green. Under the broader definition, the share of green products and services was substantially larger, but still constituted only a relatively small part of the economy.

The services sector accounted for roughly three-quarters of green business activity; manufacturing accounted for about 13 per cent; construction and agriculture made up the remaining share.

Energy conservation, resource conservation and pollution control accounted for the predominant share of green business activities, making up about 80 to 90 per cent of green shipments/receipts and employment. Renewable/alternative energy and environmental assessments accounted for a smaller share of green products and services shipments/receipts and employment.

Between 2002 and 2007, the share of green shipments and green jobs in manufacturing remained fairly constant, ranging between 0.9 per cent and 1.3 per cent. Green manufacturing jobs fell over this period, as did the jobs in all manufacturing.

The Green Growth Group (2013: 31) estimates that the United States had the largest share (19 per cent) of the global low carbon and environment market, worth nearly €800 billion.

Table 2.2 Size of green economy by industry sector in the United States, 2007

Industry sector	*USD billions*		*Share of total industry sector (%)*	
	Narrow	*Broad*	*Narrow*	*Broad*
Agriculture	2	2	0.5	0.5
Construction (green buildings and services)	36	49	2	2.8
Manufacturing	48	69	0.9	1.3
Services	286	397	1.3	1.8
All sectors	371	516	1.3	1.8

The United States ranks well in the GGEI perception survey (sixth), driven by high scores as a target for investment and green innovation, and the perception that the United States is successfully greening key sectors like buildings and transport. It performs also well in the markets and investment category, but poorly on the climate change performance measure and on many environmental categories, most notably fisheries and forests. Based on the GGEI performance index, it ranks twenty-eighth.

Japan

Transition to green economy requires industrial restructuring

In a changing economic, social and international context, Japan has made important progress towards greening its economy. It has managed to reduce some of the pressures on the environment, notably energy and material use, air emissions, water abstractions and municipal waste generation. However, complex, long-term challenges remain: primarily, climate change and biodiversity sustainable use.

In 2003, the Ministry of the Environment estimated the size of environmental business. According to this study, the eco-business turnover was JPY30 trillion (US$280 billion) in 2000.

Resource management, broadly defined and including housing renovation and repair, was estimated to account for two-thirds of the market in 2000; environmental protection, mainly waste water treatment facilities and provision of waste services, accounted for the remaining third (see Table 2.3).

It was projected that turnover would nearly double by 2020. The New Growth Strategy aims at enlarging the environment- and energy-related markets by an additional JPY50 trillion (US$530 billion).

Japan accounts for 6.3 per cent of the global market for EGS, including renewable energy technologies and low carbon activities, and in 2007–2008 was an estimated US$1.6 trillion. This is the third largest share after the United States and China.

In the second half of 2008, overall Japanese exports suffered from declining global demand, but exports of environmental products grew by over 35 per cent compared to the same period in 2007.

According to a survey conducted by the Japanese External Trade Organization in 2009, some 18 per cent of Japanese manufacturing businesses were producing and exporting environmental products, especially waste treatment devices, eco-paints and adhesives, PV cells and other renewable energy technologies, electric, hybrid and fuel cell vehicles, and waste water treatment equipment.

Overall, the transition to a green economy will require industrial restructuring.

In the Ministry of the Environment study, related employment was estimated at 769,000 in 2000, equivalent to 1.2 per cent of total employment, and was expected to grow by 46 per cent in 2010 and reach 1.2 million employed by

Table 2.3 Market sizes and employment potential of the EGS sector in Japan

	Market size (JPY billion)			*Employed*		
	2000	*2010*	*2020*	*2000*	*2010*	*2020*
Pollution management, of which:	9,594	17,943	23,706	295,570	460,479	522,201
Air pollution control	642	3,166	5,169	8,971	39,306	53,579
Waste water management	4,818	5,821	5,831	59,099	62,353	54,224
Waste management	3,614	7,736	11,126	211,859	330,006	378,035
Cleaner technologies and products	174	453	609	3,108	10,821	13,340
Resources management, of which:	20,177	28,830	34,061	468,917	648,043	700,898
Recycled materials	7,898	8,744	9,404	201,691	211,939	219,061
Renewable energy	163	929	929	5,799	30,449	28,581
Energy conservation and energy management	727	48,883	7,868	13,061	160,806	231,701
Other	10,794	13,720	15,275	218,436	219,059	195,655
Total	29,944	47,227	57,376	768,595	1,119,243	1,236,439

Source: Capozza, 2011: 38.

2020. The largest growth in employment and market value was expected in the energy sector and in the manufacturing sector for air pollution control equipment (Table 2.3).

The growth is faster than expected: Data for 2006 show that employment in private environment-related businesses had already reached 1.4 million in 2006, compared to about 76,500 employed in public environmental administration (OECD, 2010). While it was estimated that in 2010 the environmental sector would reach JPY47 trillion (Table 2.3), it reached JPY80 million (Chapter 1, Table 1.7).

As a major development in Japan towards a green economy, a policy package was adopted in 2010 that comprised of:

- Biodiversity National Strategy.
- Draft Bill of Basic Law on Climate Change.
- New Growth Strategy.
- Mid-and Long-Term Roadmap for Global Warming Measures (Draft), and
- Vision of Economic Growth by Environmental Policy.

According to the 2014 GGEI, the perception of Japan's green economic performance (seventh) dramatically exceeds GGEI performance (forty-third). Japan experienced significant improvement in its perception survey results in 2014 as compared to 2012. In the top ten overall, Japan is perceived as particularly strong on efficiency sectors and markets and investments anchored by a perceived commitment to clean transport and clean-tech innovation. Yet these perceptions are not confirmed by Japan's performance results. In that regard, Japan ranks overall towards the bottom of OECD nations. A relatively carbon intensive economy negatively impacts Japan's performance on the efficiency sectors dimension, despite better performance results for markets and investment and environment and natural capital.

Despite phasing out nuclear power, and 1.5 per cent real GDP growth, in the 2012–2013 period, Japan managed to decrease energy related emissions by 0.8 per cent and carbon intensity by 2.3 per cent (PWC, 2013; also see Table 2.1), thus decarbonising (or greening) the economy.

Australia

Sharp decrease in global green economy ranking in 2014

Australia's economy is one of the most carbon-intensive in the world. Its per capita emissions are nearly twice the OECD average and more than four times the world average, indicating significant greening potential.

The Australian Conservation Foundation estimated that in 2007 the value of Australian green industries was AUD15.5 billion (ACTU, 2008). Its growth potential until 2030 is estimated at AUD243 billion. This can be regarded as a conservative estimate, as the size of Australian environment protection industries in 2004–2005 was estimated at AUD15–20 billion, and Australia's green economy potential for 2009 was estimated at AUD45 billion – including the revenue of the environmental science service industry (AUD5 billion), waste disposal (AUD8 billion), recycling sector (AUD11 billion), clean-tech companies (AUD9 billion) and eco-tourism (AUD12 billion) (Eltham, 2010).

The Great Barrier Reef World Heritage Area[6] alone adds AUD5.68 billion (US$5.10 billion) a year to the Australian economy and generates almost 69,000 full-time equivalent jobs. Despite its significant economic and environmental value, the reef's coral cover has halved since 1985, mostly due to outbreaks of the coral-eating crown-of-thorns starfish, which are fuelled by fertiliser run-off from farms. Thus, the World Heritage Committee is considering adding the Great Barrier Reef to its 'In Danger' list.[7] The introduction of new practices that reduce pollution and improve farm productivity have been supported by AUD750 million (US$670 million) over the last ten years by Australian national and state governments and led to implementing better practices by sugarcane growers who have reduced pesticide pollution by 15 per cent and fertiliser pollution by 13 per cent. The achievements in greening protected areas indicates even higher potential for greening economic practices – in agriculture, water, waste management and recycling, and energy sectors, as well as in cities and urban transport. The potential can be illustrated by the fact that Australia's ecological footprint is nearly three times the average global footprint (WWF, 2014).

Between 2001 and 2009, increases in water prices combined with water policy reform have led to significant increases in water efficiency and high levels of water recycling. Over this period, Australia also doubled water resource productivity. That is, the Australian economy, by 2009 gained twice as much economic value from each drop of water consumed as it did in 2001.

Leading Australian businesses have achieved 70–80 per cent reductions in freshwater potable water usage, often within five years, without compromising business profitability. Australia now has a globally competitive water services sector that is highly successful.

In the energy sector, policy reforms have led to a decoupling of GDP from electricity consumption and GHG emissions. Still, overall the Australian GDP growth is highly coupled to increasing overall resource usage and consumption, and biodiversity loss and environmental degradation. Australia has made progress on reducing pollution but not to the same degree as current world's best practice. Thus, it is expected that energy, resource and environmental efficiency will be key drivers of productivity into the future. Increased efficiency would reduce infrastructure costs and the price of energy and water delivery.

Based on GGEI in 2013 Australia came second out of 27 countries for political leadership, and tenth overall for its green economic performance.

In 2014, Australia has fallen sharply in international green economy rankings. It is ranked last out of 60 countries for performance on political leadership and climate change, second for environmental and natural capital; eighteenth for investment and innovation, thirty-sixth for efficient resource use and thirty-seventh overall. In perception assessments, Australia ranked eleventh overall, seventh for environment, tenth for efficiency, twelfth for investment and eighteenth for leadership.

The 2014 index included more countries and indicators than previous editions, but Australia's fall in international standings was nonetheless 'dramatic'.

Australia is among countries that get more credit for their green economic performance than they deserve. According to the 2014 GGEI, perception of Australia's green economic performance (eleventh) dramatically exceeds their GGEI performance (thirty-seventh).

Canada

The economy is greening, but still far from being green

There is no consensus about the definition of green economy in Canada. The Canadian Institute for Environmental Law and Policy (CIELAP) conducted a national research study between November 2010 and February 2011 to seek the perspectives of diverse Canadian stakeholders on what a green economy could mean for Canada (Web and Esakin, 2011).

While all those who were interviewed had given thought to green economy concepts and many of their institutions were applying green economy ideas, very few were using the term 'green economy'. It was identified that Canadians approach green economy concepts from different perspectives, such as the concern over climate change, an interest in leveraging environmental and economic gains or a window of opportunity for innovation. There is a prevailing view that the green economy is not separate from the main economy, but greener. Thus, a green economy could be used as a unifying concept to link environmental, economic and social objectives.

Various Canadian definitions that have been put forward are largely focused on identifying and quantifying EGS, as opposed to defining it in a way that includes the rest of the economy. This gap is particularly acute on the question of what the definition of a green economy is – which makes sense given Canada's national circumstances – and how that green economy might be measured and monitored at both a macro- and micro-economic level.

Without such a definition and indicators, decision makers in both the public and private sectors are not equipped to properly plan and undertake initiatives that contribute to the transformation to a green economy. Moreover, this limited definition creates a false perspective on the role that other sectors of the economy currently play in the green economy, and limits overall understanding of how the economy as a whole can move on to a more sustainable path.

While there were different views on what the green economy is, it is perceived as a tool that could make significant gains. About two-thirds of those interviewed consider that Canada could increase its competitive advantage and achieve other benefits, including: jobs; major cost savings; attracting stable and long-term foreign investment; increased job satisfaction and retention; greater business productivity; earning revenues from improved ecosystems; and local economic stimulation.

Also, being a leader in a green economy is considered as potential for significant gains in reputation.

The main features of green economy were identified by a survey conducted by ECO Canada (2010). The results revealed that green companies were more likely to provide services (professional, scientific and technical), as opposed to products. Half of green companies are small (employing five employees or less). Among companies that consider themselves green (mostly because they provide services, or work with technology or materials that reduce environmental impact), about 14 per cent operated outside the parameters of a green company as defined for the purpose of the ECO Canada study. These parameters include three main definitional criteria, i.e. an economic perspective, a technical perspective and a development process perspective.

The economic perspective relates the characteristics of an activity relevant for its categorisation according to the economic classification system of sectors, industries, and occupations. Economic criteria might assess whether products or services contribute to decreased GHG emissions, or include sustainable resources in manufacturing processes.

The technical perspective defines the green economy through the application criteria that measure exactly what it is about a product, process or service that is 'green' and to what extent.

The development process perspective identifies where in the development cycle a green job is situated. The development process includes the phases of the development of a product or service, from the research phase to design, delivery, implementation, ongoing use and maintenance (ECO Canada, 2010).

The quantitative research conducted by ECO Canada excluded 14 per cent of the organisations that were stated green as per the terms of the working definition. The primary reasons for these exclusions were related to:

- Economic perspective (the organisations did not operate within a NAICS six-digit code associated with the green economy; and
- Technical perspective (the organisations did not produce either goods or services designed to minimise environmental impact).

The available analyses (e.g. ECO Canada 2010, Webb and Esakin, 2011, Sustainable Prosperity, 2012) suggest that the Canadian economy is *greening*, but still far from *being green*. There is no sufficient data/indicators that could support policy makers in developing policies, setting priorities and tracking progress. Measures that could foster greening the Canadian economy include phasing out subsidies to the energy and mining sectors, the stimulation of the application of new technologies and eco-industry and the implementation of fiscal incentives through carbon or energy taxation.

Based on the GGEI Canada's results are difficult to decipher. On the one hand, Canada's green economy is quite strong, both in terms of perception, where it ranks twelfth overall, and performance (twenty-ninth), where Canada does very well on the efficiency sectors and markets and investment dimensions.

But Canada's performance score near the bottom of the leadership and climate change dimension drags down its overall result significantly. The country's head of state, media coverage, behaviour in international forums and climate change performance are all extremely low as measured by the GGEI.

Mexico

Green economy programme addresses concerns for poverty alleviation and job creation

Mexico recently adopted several measures aimed at greening its economy: it launched a special climate change programme, which provides ambitious short-term targets for the reduction of GHG emissions, adopted a national strategy on sustainable consumption and production and started using economic and environmental accounts and established the Sustainable Development Centre.

The main challenges in socio-economic policy and environmental policy are:

1. Capacity building in state and local governments for effective implementation of a green economy programme, with an emphasis on poverty reduction strategies; and
2. Increasing resources for environmental policy implementation and enforcement.

Therefore, the main opportunities for a green economy programme are:

1. Improving the implementation of environmental regulation, and address concerns for poverty alleviation and job creation. This relates to improving inspection regimes and effectively extending the use of economic instruments.
2. Broadening the financial base for environmental policies and environmental fiscal reform (i.e. promoting environmental fiscal reform).
3. Improving the competitiveness of SMEs in green economy sectors (phasing out subsidies to energy and mining sectors, stimulating eco-innovation and encouraging the development of the eco-industrial sector, including the field of renewable energy) (Maradan *et al.*, 2011).

In 2012, in partnership with UNEP, the Green Economy Scoping Study (GESS) was launched. GESS provides an assessment of existing efforts related to sustainable development and proposes policy options that will encourage investment in selected key sectors, with the aim of stimulating inclusive growth, creating green jobs, and improving environmental sustainability and economic competitiveness.

According to the Green Growth Group (2013) the low carbon and environmental market in Mexico was worth over €80 billion annually and has grown by over 20 per cent in the last four years. The clean energy investments increased fivefold in 2012 and exceeded US$2 billion in 2012.

The ranking according to perception and performance components of the GGEI are basically equal (twenty-eighth perception, thirty-first performance), and the GGEI reveals various opportunities for improvement in both areas. The perception of Mexico's leadership performance has fallen significantly in 2014 compared to 2013 and performance in other areas is mediocre (the efficiency, markets and investment, and environmental and natural capital dimensions). Mexico is ranked third in performance in sustainable tourism.

Brazil

Developed market for clean-tech commercialisation

Brazil is the fourth largest GHG emitter in the world, but also one of the countries with the largest potential to reduce emissions. The green economy potential of Brazil is defined by its current economic structure and environmental conditions and natural resources abundance. Forest represents 56 per cent of national territory and 10 per cent of global forests. Brazil has approximately 12 per cent of the surface water available on the planet.

In 2012, Brazil had the seventh largest share of the global low carbon and environmental market, worth over €120 billion. It grew by nearly 24 per cent over the last 4 years. Clean energy increased by 10 per cent in 2011 and was worth US$8 billion (Green for Growth Group, 2013).

Measured by the global green energy index, Brazil's overall perception and performance scores are nearly even (fifteenth and eighteenth respectively). Brazil performs well on the markets and investment dimension: it has a developed market for clean-tech commercialisation and an attractive domestic climate for investment in renewables. Brazil has significant natural resources and its economic power is rising. Thus, the improvement of leadership and performance related to environment and natural capital (particularly forests and water) is expected to contribute to greening the economy.

Republic of Korea

While 2 per cent of GDP is allocated to improve energy security and cut emissions, no green country image

In the 2008–2010 period, around 30 South Korean business groups had invested more than KRW15.1 trillion (around US$13.6 billion) in the green sector, starting with KRW2.4 trillion (US$2.2 billion) in 2008, KRW5.4 trillion (US$4.9 billion) in 2009 and KRW7.3 trillion (US$6.6 billion) in 2010. These companies planned to invest another KRW22.4 trillion (US$20.1 billion) between 2011 and 2013.

Since South Korean Green Growth's announcement in 2008, Korean companies with technologies in the new renewable energy (NRE) sector have grown

2.2 times in terms of their total number, 3.6 times in the size of employment, 6.5 times in sales, 5.9 times in exports, and 5 times in terms of private sector investment. It is estimated that the expansion of NRE business created about 27,000 additional jobs in 2012, and is expected to create 110,000 jobs in 2015. NRE's export income is expected to reach US$36 billion in 2015 (compared to US$2.6 billion in 2009).

South Korean public and private sectors are making considerable investments in greening economic growth through the development and commercialisation of NRE technologies. The amount of investment in green growth in the year 2011 is calculated to be 3 per cent of Korean GDP or approximately US$33 billion, which is three times higher than the amount recommended by the UN.

Comparison of this amount of investment with income from export (about US$10.7 billion in 2012) shows that green growth is not very effective. It significantly relies on the domestic low carbon and environmental market, which was in 2012 worth over €70 billion annually, and has grown by over 20 per cent since 2009. The investment in clean energy reached US$2 billion, which is a more than fivefold increase compared to 2011. Further efforts to green the economy are planned, including the allocation of 2 per cent of GDP to cut emissions, improve energy security and promote new low carbon growth (Green for Growth Group, 2013). Korea aims to increase its global market share of green technology exports from 2 per cent in 2009 to 10 per cent by 2020 (World Bank, 2012).

On the other hand, based on the GGEI, South Korea does not register as a green country brand – based on performance it is ranked thirty-ninth out of 60 on the 2014 index, and on perception twenty-third. South Korea's GGEI performance result is driven by the relative carbon inefficiency of its economy and poor environmental performance, as it performed well according to the markets and investment dimension.

Turkey

Environmental issues are not a major concern of the citizens

In the early twenty-first century the growth became more inclusive, as a result of broad-based industrial development and massive job creation throughout the country. Income gaps between regions have narrowed and relative income poverty in all Turkey's 12 regions declined between 2006 and 2012. Despite the progress in areas related to education, work/life balance, environmental quality and subjective well-being, more efforts are needed. As regards air quality, PM10 levels (particulate matter that can be inhaled into the deepest part of the lung) are 36.7 $\mu g/m^3$ (with higher peaks in certain regions), far above the guideline limit of 20 by the World Health Organization. Only 61 per cent of people say that they are satisfied with water quality, the lowest in the OECD (the OECD average is 84 per cent). Some 33 per cent of people feel that they have no access to green spaces (OECD average is 12 per cent) (OECD, 2014).

Despite significant environmental challenges, the survey conducted in 2011 shows that a minority of people in Turkey believe that environmental issues are a major concern in Turkey (*Hürriyet Daily News*, 2011).

Turkey has made considerable progress in terms of greening its economy since the 1990s, especially at the political and legislative level. The implementation can be further accelerated in energy, transport, agriculture, industry and services. These sectors create pressure on the environment, use natural resources intensively, have high competitiveness and export capacity, support employment and have interconnection with SMEs, and are influenced by the approximation process to EU environment legislation.

The need to green the energy sector is underlined by increasing demand and import dependency (of about 70 per cent). Greening the energy sector is expected to contribute to raising competitiveness while preventing environmental degradation. The measures include:

- increasing the share of renewable energy (in total electricity generation to 30 per cent until 2023);
- reducing energy intensity (at least 20 per cent in 2023 compared to 2011);
- saving 20 per cent energy (improving energy efficiency in public buildings and facilities until 2023);
- saving energy in industries and energy production (by increasing the combustion efficiency of existing coal-fired thermal power plants).

In the transport sector, increasing the share of railways (up to 15 per cent in cargo transportation and 10 per cent in passenger transportation as of 2023), while decreasing the share of highways (below 60 per cent in cargo transportation and 72 per cent in passenger transportation as of 2023) contributes to the reduction of the emissions caused by the use of private vehicles. The optimisation of flight routes and modernisation of air traffic systems will lead to an increase in energy efficiency and the minimisation of environmental effects.

Agriculture depends on biological diversity and natural resources (water, soil). Degradation of these resources negatively affects agricultural activities. Greening the agriculture sector prevents income, productivity and employment loss, especially considering climate change that affects water sources and biological diversity in Turkey. Greening the agriculture sector requires protection of genetic resources, carbon capture areas (pastures, forests and agricultural lands), effective and efficient use of agricultural land and water resource, as well as measures in the forestry sector (extend forestry certification, decrease the pressures on natural forests by extending industrial afforestation, increase income generation from forestry products and services).

Greening the industry sector requires the renewal of existing technologies and the adoption of clean technologies in new enterprises in order to increase resource efficiency, and to reduce the pressures on environment and natural resources.

The services sector provides an opportunity for greening the economy as its labour oriented structure creates less pressure on the environment. The development of subsectors such as health, education, urban infrastructure, tourism, finance, information technology and consultancy includes applications towards sustainable development.

While there is significant greening potential, and progress at the legislative level, implementation is still partial. Turkey adopted the Climate Change Action Plan (CCAP) in 2011, which covers actions in energy, buildings, industry, transportation, waste, agriculture, land use, forestry and the adaptation to climate change between the year 2011 and 2023. While the plan might be used for greening the economy, it does not contain any emission reduction targets. Further, the planned measures are not being implemented.

Thus, it is concluded that Turkey enacts a carbon oriented growth model and has not adopted specific emissions reduction (or decarbonisation/greening) goals (Algedik, 2013).

Such conclusion is also supported by Turkey's GGEI, as the most concerning survey result is in the leadership and climate change dimension, where Turkey ranks last of the 60 nations polled. Turkey has a respectable performance score in terms of its attractiveness for renewable energy investment. It also performs well in terms of national efforts to facilitate this investment.

China

2–4 per cent of GDP a year to clean up 30 years of industrial waste

The strategic goal of China is to boost the country's economic growth while also enhancing environmental protection. China seeks to transform its growth pattern and engage in greener, more sustainable development.[8]

Namely, during the last 30 years China's GDP has grown by an average of 10 per cent annually, which is among the fastest rates in the world.[9]

This rapid industrial development had high environmental cost – especially regarding air and water quality, but also caused land pollution, desertification and the loss of biological diversity.

It is estimated that China needs to spend between 2 and 4 per cent of its GDP a year to clean up 30 years of industrial waste (Sim and Rong, 2010). The environmental challenges point to the need for a sweeping revision of China's development paradigm, and many Chinese policies and actions – both ongoing and planned – are in line with the green growth paradigm.

Since the 2006–2010 planning period, China has prioritised green development in almost all leading economic sectors. Efforts have been undertaken to boost energy efficiency in industry, transportation and buildings; to develop wind, solar and other renewable energy sources; to create a resource-saving 'circular' economy; and to transform traditional sectors through the use of energy-efficient and environmentally sound technologies.

As of the end of 2010, the total value of China's energy conservation service industry was RMB80 billion (US$12 billion). The industry reduced power consumption by 10.64 million tons of coal each year and reduced GHG emissions by 26.62 million tons of CO_2 (World Bank, 2012).

In 2012, China had a low carbon and environmental market worth over €500 billion (13 per cent of the global market) and was the biggest single country investor in clean energy. It invested US$65.1 billion (a 20 per cent increase on 2011), adding 16 GW of wind and 3.2 GW of solar generating capacity that year. Clean energy capacity reached 152 GW, 23 per cent of the world total (Green Growth Group, 2013).

Yet China's green economy potential has barely been tapped. China is currently in a transition period, during which new green sectors, such as renewable energy and electric vehicles, co-exist with old 'brown' industries, such as coal-based power. One of the greatest promises of China's green transition is the potential for expanded employment in industries and economic sectors, with the potential to reduce or at least slow the country's environmental impact. China's green economy plans are focused in three broad target areas: boosting the share of renewable energy in the country's energy supply; preserving and enhancing the country's natural assets, particularly forests; and embracing new technologies to reduce the environmental impacts of traditionally 'dirty' sectors. The wind power industry alone could account for over US$25 billion per year in investment, assuming 20 GW installed per year. If the State Council targets are met, the contribution of emerging green industries to China's GDP will be 15 per cent by 2020 (World Bank, 2012). By 2030, the projected exports of green technologies and services specifically related to renewable energy and clean energy (mainly electric) vehicles will rise to US$229–395 billion in export sales and 4.4–7.8 million new jobs. These export sales are on the order of 6–10 per cent of total projected exports, or 2–3 per cent of projected GDP.

Greening can and should occur in all sectors of the economy. This requires:

- Supporting better management of natural resources: air pollution, water, deforestation and desertification.
- Improving the environmental policy framework, implementation and enforcement.
- Developing and investing in sustainable buildings, stimulating further international competition.
- The further development of full cost accounting conventions inclusive of externalities to be used in energy development projects.

According to the GGEI, China ranks near the bottom of the performance measure (fifty-fifth out of 60 countries), driven by its poor performance on efficiency sectors and environment and natural capital. China ranks quite well (thirteenth) on the perception survey. China's best results in terms of the perception survey relate to the markets and investment dimension, as it is perceived as a leading target for investment and green product development along with the

United States and Germany. China ranks last in the environment and natural capital dimension.

Final remarks

The term 'green economy' has become increasingly common but without a common definition. In line with the concept of green jobs, some definitions refer only to the EGS sectors. These sectors are only a small subset of the entire economy (e.g. 1–2 per cent in the United States), thus such an approach neglects most economic activity and ignores the scale of the environmental challenges.

At the same time, some definitions have recognised that these 'green' sectors provide goods and services to the rest of the economy. An expanded perspective on the green economy encompasses the entire supply chain.

A broader definition of the green economy refers to an economy that minimises environmental impacts while not sacrificing economic growth and prosperity. This definition tends to favour a service- and knowledge-based economy and is sometimes called a 'clean' or 'low carbon' economy. Service- and knowledge-based economies develop demand (and the market) for green products, both domestically and abroad. Also, they recognise that brown industries can be turned into green and sustainable ones, provided that the necessary conditions are met (available technology, capital for investments in the infrastructure, the capacity building of institutions, new knowledge and skills for green jobs as well as adequate support).

The central challenge to achieving a green economy is how to manage economic growth and development within environmental constraints. The nature and urgency of those constraints vary depending on the particular environmental/national condition (e.g. air, land, water, waste).

Fundamentally, the concept of a green economy needs to be rooted in the larger context of economic growth. Thus, in the next chapter green growth is discussed.

Notes

1 UN Environment Management Group is a system-wide coordination body of over 40 specialised agencies, programmes and organs of the United Nations.

2 The performance index is developed by Dualcitizen Inc. and published annually since 2010. It is defined by 32 underlying indicators and datasets, each contained within 1 of the 4 main dimensions of leadership and climate change, efficiency sectors, markets and investment and environment and natural capital. More information, online, available at: www.dualcitizeninc.com.

3 Despite the formal role and efforts of EEA, there is no consensus about the most relevant sectors for greening the economy, e.g. according to Mazza and Binck (2012) ten sectors are identified as key for a transition to a green economy: agriculture, buildings, energy supply, fisheries, forestry, industry, tourism, transport, waste management and, water.

4 Based on World Bank's five-year average growth rates, online, available at: http://data.worldbank.org/indicator/NY.GDP.PCAP.KD.ZG/.

5 The methodology for GGEI draws from guidelines published through the OECD Handbook on Constructing Composite Indicators. The GGEI draws from a wide range of underlying datasets. Based on the mean and standard deviation for each indicator or dataset, a z-score and associated percentiles are calculated. Then, these percentile values can be aggregated in a uniform manner, generating a country score that is expressed on the spectrum of 0–100. For a more detailed explanation of the methodology behind the 2014 GGEI, see Dual Citizen website, online, available at: www.dualcitizeninc.com.
6 The Great Barrier Reef is the world's largest coral reef ecosystem and a World Heritage Site. The Great Barrier Reef is recognised as one of the most significant sites for biodiversity, supporting tens of thousands of species, many of which are of global conservation significance.
7 The United Nations world heritage body UNESCO in 2014 postponed a decision for 12 months on whether to list the Great Barrier Reef as 'in danger', giving the Queensland government time to put in place additional protection.
8 At the 2010 APEC Summit in Yokohama, Japan, President Hu affirmed that China seeks to transform its growth pattern and to engage in greener, more sustainable development.
9 Based on World Bank's five-year average growth rates, online, available at: http://data.worldbank.org/indicator/NY.GDP.PCAP.KD.ZG/.

References

Algedik, O., 2013. *Climate Change Action Plan Assessment Report.* Association for the Protection of Consumers and the Climate. Online, available at: www.onderalgedik.com/wp-content/uploads/2013/11/CCAP-AssessmentReport.pdf (accessed 30 November 2014).

Analytical Advisors, 2011. *The 2011 Canadian Clean Technology Industry Report: Selected Facts.* Online, available at: www.analytica-advisors.com/sites/default/files/CTR_2011Report%20SelectedFacts.pdf (accessed 30 November 2014).

Capozza, I., 2011. 'Greening Growth in Japan', *OECD Environment Working Papers*, No. 28, OECD Publishing. Online, available at: www.oecd-ilibrary.org/environment/greening-growth-in-japan_5kggc0rpw55l-en (accessed 20 October 2014).

De Gouvello, C., 2010. *Brazil Low-carbon Country Case Study.* World Bank Group. Online, available at: http://siteresources.worldbank.org/BRAZILEXTN/Resources/Brazil_LowcarbonStudy.pdf (accessed 4 October 2014).

ECO Canada, 2010. *Defining the Green Economy.* Online, available at: www.eco.ca/pdf/Defining-the-Green-Economy-2010.pdf (accessed 2 October 2014).

EEA, 2013. *Towards a Green Economy in Europe. EU Environmental Policy Targets and Objectives 2010–2050.* Report 8/2013. Copenhagen.

Eltham, B. 2010. *Australia's Green Economy Potential.* Centre for Policy Development. CPD Occasional Paper Number 10. Sydney.

EC, 2014. *Green Action Plan for SMEs, Enabling SMEs to Turn Environmental Challenges into Business Opportunities.* Communication from the Commission to the European Parliament, the Council, the European Economic and Social Committee and the Committee of the Regions. COM (2014) 440 Final.2.7.2014. Brussels.

European Parliament and the Council, 2013. Decision No 1386/2013/EU on a *General Union Environment Action Programme to 2020 'Living Well, within the Limits of our Planet'*. 20 November 2013.

GGGI (Global Green Growth Institute), 2012. *Agreement on the Establishment of the Global Green Growth Institute.* Online, available at: http://gggi.org/wp-content/

uploads/2012/10/Agreement-on-the-Establishment-of-the-GGGI.pdf (accessed 29 September 2014).

Green Growth Group, 2013. *Going for Green Growth.* Online, available at: www.gov.uk/government/uploads/system/uploads/attachment_data/file/253029/Green_Growth_Group_Joint_Pamphlet.pdf.

Green Investment Bank, 2014. *Green Investment Bank.* Online, available at: www.greeninvestmentbank.com.

Greenskills4greenjobs, 2013. 'Background'. Online, available at: www.greenskills4greenjobs.co.uk/background/ (accessed 26 March, 2015).

Hall, M., 2014. *EU Tables 'Circular Economy' Package with Zero-landfill Goal.* Euractiv, March. Online, available at: www.euractiv.com/sections/sustainable-dev/eu-tables-circular-economy-package-zero-landifll-goal-303259 (accessed 2 November 2014).

HM Government, 2011a. *Enabling the Transition to a Green economy: Government and Business Working Together.* London, UK. Online, available at: www.gov.uk/government/uploads/system/uploads/attachment_data/file/183417/Enabling_the_transition_to_a_Green_Economy__Main_D.pdf (accessed 4 October 2014).

HM Government, 2011b. *Skills for a Green Economy, a Report on the Evidence.* London, UK. Online, available at: www.gov.uk/government/uploads/system/uploads/attachment_data/file/32373/11-1315-skills-for-a-green-economy.pdf.

HM Government, 2014. *Europe 2020: UK National Reform Programme.* Online, available at: http://ec.europa.eu/europe2020/pdf/csr2014/nrp2014_uk_en.pdf (accessed 20 August 2014).

House of Commons, 2009. *Green Jobs and Skills: Second Report of Session 2008–09.* London. Online, available at: www.publications.parliament.uk/pa/cm200910/cmselect/cmenvaud/159/159i.pdf (accessed 16 November 2014).

House of Commons and Environmental Audit Committee, 2011. *The Green Investment Bank.* Online, available at: www.publications.parliament.uk/pa/cm201011/cmselect/cmenvaud/505/505.pdf.

Hürriyet Daily News, 2011. Protecting the Environment is Low Priority for Turks. Online, available at: www.greeneconomyturkey.com/articles/protecting-the-environment-is-low-priority-for-turks-says-report (accessed 20 October 2014).

IRS, 2014.*Yearly Average Currency Exchange Rates, Translating Foreign Currency into US Dollars.* Online, available at: www.irs.gov/Individuals/International-Taxpayers/Yearly-Average-Currency-Exchange-Rates (accessed 19 November 2014).

Jones V., 2009. *The Green-collar Economy.* New York: HarperCollins.

Krugman, P., 2010. Building a Green Economy. *New York Times*, 7 April. Online, available at: www.nytimes.com/2010/04/11/magazine/11Economy-t.html?ref=magazine.

Maradan, D., Maro, P., Schram, A. and Wolters, T., 2011. *Scoping Study on the Role for the Green Economy in the European Union External Action*, Brussels: European External Action Service.

Mazza, L. and Brin, P., 2012. *Green Economy in the European Union. Supporting Briefing.* UNEP.

Millennium Institute, 2012a. *Growing Green and Decent Jobs.* ITUC. Online, available at: www.ituc-csi.org/IMG/pdf/ituc_green_jobs_summary_en_final.pdf (accessed 19 November 2014).

Millennium Institute, 2012b. *United States Country Profile*, online, available at: www.ituc-csi.org/IMG/pdf/us_country_profile_mar2012.pdf.

OECD, 1999. The Environmental Goods and Services Industry: Manual for Data Collection and Analysis, OECD, Paris.

OECD, 2009. *Declaration on Green Growth*. C/MIN(2009)5/ADD1/FINAL 25 June 2009, online, available at: www.oecd.org/env/44077822.pdf.

OECD, 2010. *Interim Report of the Green Growth Strategy: Implementing our Commitment for a Sustainable Future*. Online, available at: www.oecd.org/greengrowth/45312720.pdf (accessed 25 September 2014).

OECD, 2011. *Towards Green Growth*. Online, available at: www.oecd.org/greengrowth/48224539.pdf (accessed 24 September 2014).

OECD, 2014. *OECD Economic Survey Turkey*. Online, available at: www.oecd.org/eco/surveys/Overview_Turkey_2014.pdf (accessed 5 October 2014).

PWC, 2013. *Decarbonisation and the Economy*. Online, available at: www.pwc.nl/nl_NL/nl/forms/download/decarbonisation-and-the-economy.jhtml?processed=true (accessed 30 March 2015).

PWC, 2014. *Two Degrees of Separation: Ambition and Reality*. Low Carbon Economy Index 2014. Online, available at: www.pwc.co.uk/assets/pdf/low-carbon-economy-index-2014.pdf (accessed 30 March 2015).

SDSN and IDDRI, 2014. *Pathways to Deep Decarbonisation*. Paris: Science Po. Online, available at: http://unsdsn.org/wp-content/uploads/2014/02/DDPP_interim_2014_executive_summary.pdf (accessed 24 October 2014).

Sim, G. and Rong, F., 2010. 'China is Set to Lose 2% of GDP Cleaning Up Decades of Pollution'. *Bloomberg News* 17 September 2010. Online, available at: www.bloomberg.com/news/2010-09-16/china-set-to-lose-2-of-gdp-fighting-pollution-as-doing-nothing-costs-more.html (accessed 2 December 2014).

Sustainable Prosperity, 2012. *Towards a Green Economy in Canada*. Online, available at: www.sustainableprosperity.ca/dl864%26display.

Tamanini, J., 2014. 'The Global Green Economy Index'. *GGEI 2014 Measuring National Performance in the Green Economy 4th Edition*, online, available at: http://dualcitizeninc.com/GGEI-Report2014.pdf.

UAE Ministry of the Environment and Water, 2013. *UAE to Host First Global Conference on Partnership for Acton on Green Economy*, Statement issued by the UAE Ministry of the Environment and Water at a press conference on 17 December 2013. Online, available at: www.unep.org/greeneconomy/Portals/88/PAGE/UAE%20press%20release%20-%2017%20dec%202013.pdf (accessed 2 November 2014).

UN, 2012. *A Guidebook to Green Economy*. Online, available at: http://sustainabledevelopment.un.org/content/documents/GE%20Guidebook.pdf (accessed 6 November 2014).

UN Environment Management Group, 2011. *Working Towards a Balanced and Inclusive Green Economy (4)*. Online, available at: www.unep.ch/etb/pdf/2009%20statement%20deliver%20as%20one/Interagency%20Joint%20Statement.%20E%20rev1.pdf.

UNEP, 2011. 'Introduction, Setting the Stage for a Green Economy Transition'. Online, available at: www.unep.org/greeneconomy/Portals/88/documents/ger/1.0_Introduction.pdf (accessed 10 November 2014).

UNEP, UNCTAD, UNDESA, 2011. *Transition to a Green Economy: Benefits, Challenges and Risks from a Sustainable Development Perspective*. UN.

Webb, C. and Esakin, T., 2011. *Green Economy for Canada: Consulting with Canadians*. Canadian Institute for Environmental Law and Policy. Online, available at: http://cielap.org/pdf/CIELAP_GreenEconomy.pdf.

World Bank, 2012. *China 2030*. Online, available at: www.worldbank.org/content/dam/Worldbank/document/China-2030-complete.pdf (accessed 21 November 2014).

WWF, 2014. *Living Planet Report 2014*. Online, available at: wwf.panda.org/about_our_earth/all_publications/living_planet_report (accessed 21 November 2014).

3 Green growth

The concept

The concept of green growth has its origins in the Asia-Pacific region, which has experienced unprecedented economic growth in the last 30 years.

The concept of economic growth in Asia has been subject to various theoretical debates, including those of mainstream neoclassical observers as well as many heterodox economists who believe that the remarkable performance of Asian economies has much to do with the role of their state, rather than (or in addition to) that of market forces. The analysis of the Asian model of development was usually carried out in the framework of the national economy. On the other hand, the flying geese concept has often been used to explain the regional catching-up process, particularly after the mid-1980s. The concept is based on the implicit assumption that late industrialisers can catch up provided that they relate themselves with early industrialisers along market rationalism (Kashara, 2013).

In the period 1995–2002 in the Asia-Pacific region, demand for water, energy and raw materials increased by 50 per cent, emphasising the need for the efficient use of resources. Such approach was adopted by the Ministerial Declaration (the Seoul Initiative Network on Green Growth) and the regional implementation plan for sustainable development in 2005, when governments and other stakeholders from Asia and the Pacific agreed to move beyond the sustainable development rhetoric and pursue a path of 'green growth' (Allen and Clouth, 2012).

Based on the declaration, the United Nations Economic and Social Commission for Asia and the Pacific (UNESCAP) launched a regional initiative that supported green growth as a key strategy for achieving sustainable development and the MDGs (in particular, Goal 2 relating to poverty reduction and Goal 7 relating to environmental sustainability). The green growth approach adopted by the Ministerial Declaration sought to harmonise economic growth with environmental sustainability, while improving the eco-efficiency of economic growth and enhancing the synergies between environment and economy.

The ROK, which adopted a national green growth strategy in 2008, has promoted the green growth concept since then, including through the OECD. The OECD ministerial council, representing 30 members and five prospective members, from countries comprising approximately 80 per cent of the global economy, in 2009 approved a declaration acknowledging that green and growth can go hand-in-hand and launched the development of the OECD Green Growth Strategy. The strategy aimed to bring together economic, environmental, technological, financial and development aspects into a comprehensive framework.

In 2010 green growth was high on the agenda of various organisations:

- The Association of Southeast Asian Nations (ASEAN) adopted the ASEAN Leaders' Statement on Sustained Recovery and Development.
- The UNESCAP countries adopted the Incheon Declaration on Green Growth.
- Sixth Ministerial Conference on Environment and Development reiterated its support (originally expressed by the Seoul Initiative Network on Green Growth) for green economy by a declaration in Astana.
- G20 Seoul Summit recognised green growth as an inherent part of sustainable development; and
- The Global Green Growth Institute (GGGI) was established.

The ASEAN Leaders' Statement on Sustained Recovery and Development highlighted the leaders' determination to promote green growth, including investments in long-term environmental sustainability and the sustainable use of natural resources in order to diversify and ensure economic resilience.

The Incheon Declaration on Green Growth expressed the intent of the UNESCAP countries to 'strengthen efforts to pursue green growth strategies as part of response to the current crisis and beyond'. The main focus of green growth in this context was for developing countries in the Asia-Pacific region to harmonise economic growth with environmental sustainability, while improving the eco-efficiency of economic growth and enhancing the synergy between environment and economy.

The G20 Seoul Summit recognised green growth as an inherent part of sustainable development. Green growth was seen as a tool that could enable countries to leapfrog old technologies.

The GGGI was founded in 2010 as a non-profit foundation in Seoul to advance a green growth model. GGGI was elevated to a new international organisation at the Rio+20 Conference in 2012. It is dedicated to diffusing green growth as a new model of economic growth (targeting poverty reduction, job creation and social inclusion) and environmental sustainability (climate change mitigation, biodiversity loss, and the security of access to clean energy and water).

GGGI identifies three major barriers to green growth:

- A lack of practical experience in planning and implementing green growth strategies, and in generating demonstrated results;
- The absence of a convincing economic theory and policy agenda to explain the fundamentals of green growth and guide its pursuit;
- Insufficient engagement of the private sector, caused by slow changes in behaviour by industry and financial markets that would allow the scaled deployment of resource-efficient technologies and the internalisation of environmental externalities in capital allocation.

Thus, GGGI focuses its activities on:

- The adoption and implementation of green growth plans (with adequate local stakeholder support, financing and technologies);
- The provision of relevant, high-quality research for policymakers; and
- The engagement of the private sector in the implementation of national green growth plans.

The GGGI seeks to serve as a bridge between developed and developing countries, the public and private sectors, and practitioners and scholars, respectively. The overall goal is to channel the GGGI's knowledge-base, networks and experience to provide the best technical assistance and support to developing countries, promoting and implementing their own rigorous green growth economic development strategies. Membership to the GGGI is open to any member state of the United Nations that subscribes to the organisation's goals and objectives. Regional integration organisations – bodies consisting of sovereign states of a given region – are also eligible for GGGI membership.

In 2011 the OECD's strategy *Towards Green Growth* was launched. It defined green growth as the means for fostering economic growth and development while ensuring that natural assets continue to provide the resources and environmental services on which human well-being relies (OECD, 2011). Since then, the OECD has become a major proponent of green growth. It supports the efforts of countries to implement green growth. According to the OECD, green growth leads to a low carbon, climate resilient and resource efficient economy (OECD 2011). It is a way to pursue economic growth and development, while preventing environmental degradation, biodiversity loss and unsustainable natural resource use (OECD, 2010). The main elements of green growth are:

- green investments;
- R&D;
- low carbon infrastructure;
- tax instruments;
- the co-ordination of the labour market with education policies; and
- international cooperation (OECD, 2009, 2010).

In its 2011 report, the OECD highlighted five additional sources of economic growth from environmental policies: productivity (enhancing productivity, reducing waste and energy consumption and making resources available to highest value use), innovation, the creation of new markets and job opportunities, boosting investor confidence, and stability in macroeconomic conditions (e.g. reduced resource price volatility) (OECD, 2011).

A review of comparator indicators shows the position of selected countries and highlights the greening potential of each country. These indicators (summarised in Table 3.1. from available data) are grounded in the OECD integrated

Table 3.1 Energy and emission indicators

	Energy/capita (kg of oil equivalent)	*(PPP) GDP/ energy use (USD per kg of oil equivalent)*	*(nom.) GDP/ energy use (USD per kg of oil equivalent)*	*CO_2/capita (kg)*	*CO_2/(PPP) GDP (kg/USD)*	*CO_2/(nom.) GDP (kg/USD)*	*CO_2 growth 1990–2009 (%)*
Upper middle-income countries	2,177	5.2	n/d	5.3	0.5	n/d	n/d
Euro area/OECD Europe	3,763	8.2	5.6	8.4	0.3	0.4	−5
Eastern Europe and Central Asia	3,030	3.6	0.7	7.2	0.7	3.3	−38
World	1,835	5.5	3.3	4.6	0.5	0.7	38
Republic of Korea	5,260	6.1		11.5	0.4		80
Japan	3,539	9.9		9.2	0.3		−3
Canada	7,270	5.7		14.7	0.4		−5
Australia	5,883	7.2		16.9	0.4		8
EU	3,482	9.9		7.9	0.3		
Turkey	1,333	8.9	3.6	4.0	0.3	0.7	102
Brazil	1,295	7.4	3.6	1.9	0.2	0.4	74
China	1,598	3.6	1.4	5.0	0.9	2.2	207
India	545	5.1	1.3	1.4	0.5	1.8	172
Mexico	1,698	7.9	4.1	4.5	0.3	0.6	51
Russia	4,838	3.1	0.6	10.8	0.8	3.9	−30
United States	7,503	5.8	5.3	19.3	0.4	0.5	7

Sources: 'World Development Indicators', 2011 edition, the World Bank; GDP in 2005 PPP USD; except nominal GDP/energy use, CO_2/nominal GDP and CO_2 growth per cent 1990–2009 (sectoral approach) from 'CO_2 Emissions From Fuel Combustion', 2011 edition, the International Energy Agency; GDP in 2000 USD. World Bank figures/Euro area, IEA figures/OECD Europe.

framework for monitoring green growth progress. The indicators are grouped into four information themes: environmental productivity, resource productivity, natural asset base and life quality.

On the other hand, according to a recent report of the Institute for Sustainable Development and International Relations (IDDRI),[1] green growth relies basically on decarbonisation (IDDRI, 2014), which requires:

- Energy efficiency and conservation, particularly in transport, buildings and industry;
- Low carbon electricity, with the replacement of fossil fuels with nuclear or renewable energies, such as solar, hydro, wind and geothermal, or the coupling of fossil fuel combustion with CCS; and
- Switching end-use energy supplies from fossil fuels in transport, buildings and industry with lower carbon fuels (IDDRI, 2014).

Not only are the main drivers and barriers of green growth are identified, there are at least 13 separate definitions for green growth identified in recent publications, which include definitions by key international actors involved in green growth work, such as UNESCAP, OECD, the World Bank and GGGI (see Box 3.1).

Box 3.1 Selected definitions of green growth

1 Aims to foster economic growth and development while ensuring that natural assets and environmental services are protected and maintained. The approach places a premium on technology and innovation – from smart grid systems and high-efficiency lighting systems to renewable energies including solar and geothermal power – as well as on improving incentives for technology development and innovation (Global Sustainability Panel, 2011).

2 Fosters economic growth and development, while ensuring that natural assets continue to provide the resources and environmental services on which our well-being relies (OECD, 2011).

3 A policy focus for the Asia-Pacific region that emphasises environmentally sustainable economic progress to foster low-carbon, socially inclusive development (UNESCAP website).

4 Is, in general terms, economic progress that fosters environmentally sustainable, low carbon and socially inclusive development. Pursuing green growth involves outlining a path to achieving economic growth and well-being while using fewer resources and generating fewer emissions in meeting demands for food production, transport, construction and housing, and energy (UNESCAP, 2012).

5 Is about making growth processes resource-efficient, cleaner and more resilient without necessarily slowing them. Development that is green (which here means resources-efficient), clean and resilient (World Bank, 2011).

6 Is the new revolutionary development paradigm that sustains economic growth while at the same time ensuring climatic and environmental sustainability. It focuses on addressing the root causes of these challenges while

ensuring the creation of the necessary channels for resource distribution and access to basic commodities for the impoverished (GGGI website).

7 Green growth is growth that emphasises environmentally sustainable economic progress to foster low carbon, socially inclusive development. The OECD definition is similar but emphasises also green investment as a driver for economic growth (UNDESA – Rio+20 Objectives and Themes of the Conference).

8 It is growth that is efficient in its use of natural resources, clean in that it minimises pollution and environmental impacts and resilient in that it accounts for natural hazards (World Bank, 2012a).

9 It is growth that is efficient in its use of natural resources, clean in that it minimises pollution and environmental impacts, and resilient in that it accounts for natural hazards and the role of environmental management and natural capital in preventing physical disasters. And this growth needs to be inclusive. Inclusive green growth aims to operationalise sustainable development by reconciling developing countries' urgent need for rapid growth and poverty alleviation with the need to avoid irreversible and costly environmental damage (World Bank, 2012b).

10 Growth achieved by saving and using energy and resources efficiently to reduce climate change and damage to the environment, securing new growth engines through the R&D of green technology, creating new job opportunities, and achieving harmony between the economy and environment (RoK Framework Act on Low Carbon, Green Growth, 2010).

11 Defined as an environmentally sustainable progress that fosters low carbon, socially inclusive development (Government of Cambodia, 2009).

12 An emerging concept that recognises that environmental protection is a driver of global and national economic development. It refocuses society on achieving qualitative growth rather than simply increasing GDP (Government of Rwanda, 2011).

13 Means 'job creation or GDP growth compatible with or driven by actions to reduce greenhouse gases' (Green Growth Leaders, 2011).

Source: Allen and Clouth, 2012

As the snapshot of various definitions presented in Box 3.1 shows, the concept of the green economy is also evolving in various World Bank documents. Its main features – that green growth is efficient, clean and resilient – are unchanged, but recent definitions explain it in more detail. This especially concerns the concept of resilient development. Later explanations have added that green growth is resilient in that it accounts for natural hazards and highlights the role of environmental management and natural capital in preventing physical disasters (see Box 3.1). The green growth model is also incorporated in the World Bank Group's Environment Strategy 2012–2022, which lays out an agenda to support a 'green, clean, resilient' development path. In this context, green refers to sustainable management and the conservation of natural resources (oceans, land and forests) to improve livelihoods and ensure food security. Clean means a low-pollution, low-emission world in which cleaner air, water and oceans enable people to lead healthy, productive lives. Such growth requires innovative financing for renewable energies, climate-smart agriculture and lower-carbon cities. Resilient means prepared for shocks and adapting effectively to climate change. The World Bank stipulates the link between growth and development, and claims that the green growth model should enable sustainable, inclusive development, satisfy the basic needs of nine billion people and provide them with equal rights to material prosperity.

According to the World Bank's report *Inclusive Green Growth: the Pathway to Sustainable Development*, at the heart of green growth is the underlying assumption that environmental assets are not used efficiently (World Bank, 2012a). Therein lies the potential for green growth – growth that is efficient, clean and resilient. Green growth policies aim to foster sustainable development by reconciling the need for environmental sustainability with that for economic growth and social improvement.

In order to identify and address major knowledge gaps in green growth theory and practice, the Mexican presidency of the G20 in 2012 introduced 'inclusive green growth' as a cross-cutting priority on the G20 development agenda, and a new international knowledge-sharing platform was launched in Mexico – the Green Growth Knowledge Platform.

The Green Growth Knowledge Platform brings together the major international organisations supporting and promoting both green growth and green economy (the World Bank, UNEP, OECD and the GGGI). Its aim is to help countries to design and implement policies to move towards a green economy.

The need to finance the implementation of a new low carbon model for growth as the world rebuilds from the global economic crisis was recognised by the International Monetary Fund (IMF), which is working on proposals to create a Green Fund to finance this shift in the global economy with up to US$100 billion a year, in order for countries to confront the challenges posed by climate change (IMF, 2010).

The recovery of the largest economies can be facilitated by the transition to a green economy. On the other hand, while according to the World Bank high-income countries 'appear to be finally emerging from the global financial crisis', the panic outflow of investors from the emerging economies may further damage growth and jobs and contribute to the continuing imbalance and fragility of the global economy.

Widening income disparities and structural unemployment are the most serious problems confronting the global economy (WEF, 2014a, cf. Piketty, 2014). The current policies of most governments do little to address either of these problems. The shift in policy from austerity to the following would enable job creation and support green growth (ITUC, 2014b):

- Targeted investments in infrastructure to improve long-term productive potential and move to a low carbon economy.
- Raising the purchasing power of low- and middle-income households by reducing inequality and strengthening collective bargaining and minimum wages.
- Investing in active labour market policies to raise skill levels and reduce youth unemployment.

- Reducing informality and creating decent work in emerging and developing countries.

Green growth and deep decarbonisation are two interrelated concepts[2] that are considered to be key to determining an inclusive and sustainable future (Canzi and Tkacik, 2014). Deep decarbonisation is a radical transformation of energy systems, necessary to be achieved by 2050, including rapid declines in carbon intensity in all sectors of the economy (Sachs and Tubiana, 2014).

While there is no global consensus on how to operationalise green growth, most approaches include four main elements:

1 mitigation (of GHG emissions);
2 adaptation (to a changing climate);
3 other environmental protection (clean air and water, natural beauty and biodiversity); and
4 innovation and green jobs (the hypothesis is that a shift to green growth will spur technological innovation and promote trade competitiveness in new industries).

The relative priority put on each of these elements depends on national circumstances (Bouzaher *et al.* 2013), primarily the availability of various forms of capital.

National approaches

Republic of Korea

A first-mover in the implementation of green growth but still among the major carbon emitters

The ROK was pioneering in developing and promoting its green growth model. In 2008, partly in response to the global financial crisis, the Korean government adopted 'low carbon green growth' as the country's new development vision. It seeks to break away from the conflicting nature of 'green' and 'growth' by integrating three key principles of development:

1 maintaining productive economic activities while minimising the use of energy and resources;
2 minimising environmental pressure with each use of energy and resource; and
3 making investment on environment a driver for economic growth.

Green growth combines three mutually-reinforcing objectives:

1 responding to the economic crisis through a green stimulus;
2 reducing energy dependency; and
3 rebalancing the economy towards green sectors in the long term.

The financial crisis exposed Korea's reliance on imported energy (96 per cent of energy accounting for two-thirds of all imports) as a major weakness in its growth model. Korea aims to decrease its energy intensity by 46 per cent and increase the share of renewable energy in total primary energy from 2.4 per cent in 2007 to 11 per cent in 2030. Furthermore, the 2009–2013 five-year plan allocated 2 per cent of GDP to 10 green growth strategies, each containing quantitative objectives and well defined projects.

The National Strategy for Green Growth 2009–2030 and the Five-Year Plan for Green Growth 2009–2013 (which were accompanied by the enactment of a Framework Act on Low Carbon Green Growth) followed in 2009, which should enable action in response to climate change, the eradication of poverty and minimisation of environmental damage. Since these policies were announced, Korea has been promoting the concept of green growth more broadly, including through the OECD.

In 2013, the new government distanced itself from its predecessors' legacies – particularly with regard to its green growth initiative. However, Korea's president has since re-embraced the previous government's climate policies and embarked upon 'Green Growth 2.0'. The focus of Green Growth 2.0 is on emission reduction. Korea's goal is to curb the nation's GHG emissions by 30 per cent from business-as-usual levels by 2020. To that aim, the government has put in place the Target Management Scheme, which requires 470 companies and institutions to set emissions goals and pay fines of up to KRW10 million (US$9,400) if they fail to meet these goals (Hyon-hee, 2013). The programme is a prelude to a carbon trading system slated for introduction in 2015. Under the planned system, firms will only be able to emit more than an allotted figure by buying credits from those that have discharged less than their quota.

However, Korea is one of the major emitters and is not on track to meet this commitment. In 2010, emissions of GHGs increased by 9.8 per cent (to 669 million tons) compared to 2009, mainly due to a surge in power demand and industrial production.

Japan

Promoting Japanese technologies worldwide to reduce global GHG

After a prolonged period of economic stagnation in the 1990s (the so-called lost decade), in 2002 Japan's economy started recovering. The recovery was mainly export-driven, while domestic demand was weak. The 2008 global economic slowdown, coupled with a rise in the value of Japanese currency reduced the volume of Japan's exports. Thus, in order to restore growth, in 2009 Japan adopted the New Growth Strategy (which was revised in 2010) based on domestic demand, innovation and stronger economic integration of Japan in the Asia region. It identified the environment, and in particular innovation aimed at addressing environmental concerns, among sources of economic recovery and long-term growth.

The New Growth Strategy incorporates all the main elements of green growth identified by the OECD: it emphasises the linkages between environmental protection, economic growth and social change, in an attempt to move from a strictly environmental interpretation of sustainable development to a more integrated approach. Among the basic policies of Japan's New Growth Strategy is to become a leader in environment and energy thorough green innovation. The objectives by 2020 include:

- to generate a market value of over JYP 50 trillion and 1.4 million jobs in environment-related sectors; and
- to reduce global GHG emissions by at least 1,300 $MtCO_2eq$ by promoting Japanese technology worldwide.

Until the Fukushima nuclear power plant incident in 2011,[3] focus on energy security and climate change has favoured the development of renewables and the domination of nuclear power. The innovative strategy for energy and the environment adopted in 2012 and the comments on the basic energy plan by the Advisory Committee for Natural Resources and Energy concluded that the dependency on nuclear power should be decreased. Consequently, power generation from nuclear power decreased substantially in 2012 from its level in 2010, while the import of fossil fuels, especially liquefied natural gas (LNG), increased in spite of energy efficiency improvement in the end-use sector. Thus, to achieve the GHG reduction target of an 80 per cent reduction by 2050 compared to 1990 with lower nuclear dependence, it is necessary to improve energy efficiency, to reduce energy consumption and to increase the share of renewable energies. The potential of renewable energies is unevenly distributed, thus regional electricity exchange is required and related investments in interconnection capacities.

Green growth thus requires a large scale transformation in the energy system. In particular, the integration of variable renewable energy (solar PV and wind power) into the electricity system is a huge challenge. Demand-side management is identified as an effective option, but it may not be implemented by a market mechanism alone. Therefore, additional policy instruments (e.g. dynamic pricing of electricity) would be needed. Also, the pace of deploying low carbon technology is strongly influenced by public acceptance (Kainuma *et al.*, 2014).

China

Green growth should be based on green services and green consumption

During the last 30 years, China's economy has followed the evolutionary process of the 'flying geese paradigm'. The transformation of China's economic pattern can be divided into three stages, each lasting ten years:

- Stage 1 is the implementation of an export-oriented strategy. During this stage the proportion of manufacturing in the core industries (labour intensive light industries) rapidly increased.

- In Stage 2, the share of heavy chemical industry increased. The emphasis in this stage is on capital-intensive or resource-intensive heavy industry.
- Stage 3 is characterised by the introduction of a science and technology-oriented strategy.

The change of the economic structure led to increased energy consumption in the 1990s.

With the transformation of the economic pattern to a green economy, China's energy consumption per unit of GDP has more than halved, from 700 kg oil equivalent per US$1,000 GDP in 1990, to 300 kg oil equivalent per US$1,000 GDP in 2008. Despite remarkable progress, China's energy consumption per unit of GDP is twice the global average. Thus, there is a significant opportunity for reductions in energy intensity. This decline cannot depend on incremental technology change, because China's power plants are newly-built and efficient, and for energy-intensive industries the efficiency gap compared to developed countries is low (10–20 per cent). Therefore, the focal point in China is to adjust the industrial structure and change the mode of development towards reducing heavy and chemical industry, as well as reduced production of energy-consuming products like steel and cement. Namely, the industry sector is the main driver of economic growth in China and accounts for 68 per cent of total final energy consumption, and almost 71 per cent of total energy-related CO_2 emissions in 2010. A few energy-intensive industries (iron and steel, cement, synthesis ammonia, and ethylene production) consume about 50 per cent of energy use in the industrial sector. This is also closely linked to the process of urbanisation, which drives demand for construction materials (such as steel and cement). According to estimates, there may be an increase of 1 per cent in the urbanisation rate each year (Fei *et al.*, 2014).

Greening growth would require the evolution of the economic structure, reductions in energy intensity and the promotion of non-fossil fuel energy. Green growth could also address the demand of developing the national economy.

China's growth has been characterised by a high saving and investment rate in the past three decades. In the future, China will maintain its growth rate at around 7 per cent, reduce its saving and investment rate, and increase the share of consumption in its GDP. To maintain the growth rate at a relatively high level while reducing the investment rate, China needs to increase the productivity of its investment. Deep decarbonisation strategies can contribute to this gain in productivity through:

1. Shifting the structure of the economy towards less capital intensive sectors (e.g. from industrial sectors to service sectors);
2. Improving the efficiency of capital investment to produce output, especially through energy saving; and
3. Increasing the productivity of other factors, especially labour and energy (Fei *et al.*, 2014).

Green growth is expected to create new jobs: meeting the government solar, wind and hydropower targets could create about 6.8 million jobs (direct and indirect).

Based on the assumption that the economy will expand at around 7–8 per cent a year, and gradually shift away from basic industry towards services, it is expected that a further 20 million jobs will be created. Improving energy efficiency by 60 per cent would lead to about 17.3 million job losses, but the focus on low carbon and services sectors could outstrip these losses by almost 10 million (Global Climate Network, 2009).

There are three main sources of green growth in China:

1 green transformation of traditional sectors;
2 expansion of emerging green industries;
3 expansion of the service sector (World Bank, 2012b).

First, a large number of existing conventional techniques and management models can reduce energy use and emissions and improve the level of corporate profitability. The energy intensity of China's GDP, measured in terms of primary energy consumed per unit of output is more than twice that of Germany (390 tons of coal equivalent (tce) per million US dollars of output in 2009 and 167 tce per million US dollars respectively). There is a huge potential to generate financial savings in the buildings and industry sectors (estimated at US$65 billion per year by 2030, World Bank, 2012b). There are also additional 'co-benefits' of green growth, such as:

- improving local air quality and thus reducing the incidence of respiratory illness associated with air pollution;
- reducing infrastructure constraints in related sectors, such as transport and water; and
- reducing import dependence.

Cost-effective energy efficiency and renewable energy investments offer triple-win (win–win–win) outcomes by trimming production costs, mitigating emissions of GHGs, and improving public health risks in various sectors. The value of these health-related co-benefits is US$20 billion per year in 2030, on top of the direct savings of US$65 billion per year.

Second, the expansion of emerging green industries (such as solar and wind energy, together with equipment manufacturing and electric vehicle industries) stimulates the development of low carbon, resource-lite, and environmentally friendly technologies, goods, and exports. In addition, increased public awareness will help shift consumer demand towards green products.

China continues to absorb and innovate new green technologies, supported in part by government investment and policies, which might contribute to competitiveness and the country's growth in the upcoming years. The growth of China's nascent environmental protection industry demonstrates the importance of national policies in promoting the growth of green sectors.

Third, the green transformation will impact the service sector in two ways.

- it will enable the development of new green service industries, such as ecosystem services, carbon asset management services, carbon trading, and contract energy management; and
- it will support the country's intended economic rebalancing away from heavy manufacturing and towards a larger service sector.

Both trends enable reducing China's carbon footprint, as its efficiency in manufacturing may soon approach, or even surpass the levels of high-income countries.

The rising share of services in GDP will help reduce the economy's carbon intensity and the energy intensity of output (value added). Energy intensity by secondary industries in 2009 was eight times higher than for agriculture and five times that of services. Every percentage point increase in the share of services in GDP is associated with a decline in energy consumption of 1.4 percentage points (World Bank, 2012b).

Australia

Greening the growth relies primarily on significant decarbonisation and energy efficiency

Australia's GHG emissions per capita are high, due to the extensive use of coal in electricity supply, which accounts for roughly two-thirds of Australia's GHG emissions. Exports of energy, minerals and agricultural commodities play an important role in the Australian economy, with the relative importance of specific commodities changing over the decades in response to international demand.

Greening the growth could enable maintaining the economic growth (at 2.4 per cent annually on average) and prosperity, and decarbonisation by 2050, resulting in an economy nearly 150 per cent larger than today. Productivity keeps rising, with 43 per cent growth in real wages and exports growing at 3.5 per cent per annum (Jotzo *et al.*, 2014).

Green growth is not uniform across the Australian economy. It is driven by an increase in activities such as renewable energy generation and forestry, but significant reductions in primary industries (such as coal production, oil extraction, and heavy manufacturing) are expected. Greening Australian growth relies primarily on the decarbonisation of energy transformation (mainly electricity generation) combined with electrification (supplied by decarbonised electricity) and fuel switching. This could lead to a nearly 75 per cent reduction in the emissions intensity of energy use across all economic sectors. Also, industrial energy emissions could decrease by nearly 60 per cent by 2050, while the economic value added of industrial activities more than doubles. In buildings, increased energy efficiency and electrification of direct fuel use (e.g. gas for heating) and the use of decarbonised electricity could lead to the reduction of GHG emissions

Table 3.2 Job creation resulting from green growth in Australia over five years

EUR invested/ job	*New jobs created over 5 years, with 2 per cent investment of GDP*				
	Total	*Energy*	*Construction*	*Transport*	*Manufacturing*
46,726	919,106	94,035	343,395	342,931	102,610

Source: Millennium Institute (2012) and authors' calculation.

by 95 per cent by 2050. In transport, oil use for road transport could be reduced by 85 per cent between 2012 and 2050.

Progress in the promotion of green growth based on efficient environmental policy achieved in recent years was based on the introduction of a carbon price, which, together with accompanying measures, aimed to encourage investment in clean energy technologies. However, the government that took office in 2013 disbanded the Climate Commission, and the Senate is considering legislation to abolish the carbon tax, the Climate Change Authority and the Clean Energy Finance Corporation (CEFC).

It is estimated that green growth could enable a US$227 billion increase in the value of green sectors, from US$15.5 billion to US$243 billion and that green jobs employment could grow to 847,000 jobs by 2030 (ACTU, 2008). The main driver of growth is innovation, which should be reinforced and supported by direct government investment in R&D, incentives for private investment, and developing environmental knowledge and skills. The Millennium Institute estimate much stronger job creation within five years, provided that 2 per cent of GDP is invested in green growth: Based on that estimate, 919,106 new jobs could be created in a 5-year period in Australia: roughly one-third each in the construction and transport sectors, with the remainder in the manufacturing and energy sectors (Table 3.2).

EU and its member states

Is the circular economy tool for green growth?

For the EU the 'green economy' is a model that

> secures growth and development, safeguards human health and well-being, provides decent jobs, reduces inequalities and invests in, and preserves biodiversity, including the ecosystem services it provides (natural capital), for its intrinsic value and for its essential contribution to human well-being and economic prosperity.
>
> (European Parliament and the Council, 2013)

In the EU member states the green economy is the way to ensure growth, as opposed to the Asian concept that promotes green growth as a tool for the creation of a green economy.

The promotion of resource efficiency and the establishment of a 'circular economy' accompanied by job creation should enable decarbonisation and thus green growth (EC, 2014: 3; SSDN & IDDRI, 2014). Higher re-use and recycling rates and boosting environmentally friendly employment (job-rich measures) can create the framework conditions for a smooth transition 'to a green economy' attempting to 'avoid the dislocation' that could be caused by the current industrial model. The current model is linear (where one takes–makes–consumes and disposes (EC, 2014), while the green growth model is a circular one, i.e. competitiveness is defined by resource productivity.

According to the EC, a green growth model is used to depict a structural economic change driven by resource scarcity, technological change and innovation, new markets and changes in demand. The Europe 2020 strategy recognises the central role of the transition towards a green, low carbon and resource efficient economy in achieving smart, sustainable and inclusive growth (e.g. EC, 2014: 2)

A good quality environment supports many sectors in the economy. Before the crisis the environment was linked to around 21 million jobs and over a trillion euros of economic activity in Europe (GHK *et al.*, 2007). Today, the key factors for enabling green growths are the labour market and skills. In addition, environmental policies can be a source of new jobs and innovation, a driver of progress, and may ultimately help to increase the health of our economies and the well-being of our societies.

According to Jaeger *et al.* (2011), if Europe's emissions reduction target is raised from 20 per cent to 30 per cent by 2020, Europe's annual rate of economic growth may increase by up to 0.6 per cent, generating six million new jobs and boosting investment as a share of GDP from 18 per cent to 22 per cent.

The estimate for the European Union shows that about 500,000 net jobs could be created in 2014–2020 by investing 14 per cent of the total EU budget in renewable energy, nature conservation, green buildings and sustainable transport. Greening the investments is expected to increase job creation per euro threefold: a €1 billion investment is expected to create 130,000 jobs (GEF, 2013).

Sector specific studies provide different estimates: the potential for green jobs creation in 2020 in energy efficiency in buildings is estimated between 261,400 and 378,000 new jobs, in renewable energy 2.7 million (based on 20 per cent renewable target) and in the waste and recycling sector between 2.4 and three million jobs, depending on the recycling rate (50 per cent and 70 per cent respectively, see Table 3.3).

To achieve the potential for green job creation in 2020 (about six million jobs), necessary investment is about €200 billion (Table 3.3). The necessary investment for green job creation in the EU ranges between €19,000 and 152,000. There are, however, significant differences among EU member states and sectors.

The analyses made by the Millennium Institute show that necessary investment in order to create a new job in Germany is almost four times higher than in Bulgaria (Table 3.4), but the investment potential in Germany is 40 times higher than in the Bulgaria (4.3 million jobs compared to 177,000). While the cost of

Table 3.3 Sector specific potential for green job creation in the EU and necessary investment

Sector	*Number existing jobs*	*Potential for job creation 2020*	*Potential for job creation 2030*	*Potential for job creation 2050*	*€ invested/job*	*Necessary investment until 2020, billion €*
Eco industries	2,900,000–3,600,000	n/a	n/a	n/a	n/a	n/a
Energy efficiency in buildings	232,000	261,400–378,000	n/a	n/a	38,610	10.1–14.6
Renewable energy	1,114,210	2,700,000	4,400,000	6,100,000	18,957	51.2
Efficient transport, sustainable mobility	2,100,000	n/a	n/a	n/a	46,511	n/a
Organic farming	220,000–260,000	2,400,000–2,963,000	n/a	n/a	151,515	n/a
Biodiversity conservation	n/a	n/a	n/a		34,482	n/a
Waste and recycling	2,963,000	n/a	n/a	n/a	46,948	139
Total	n/a	5,924,400–6,051,000	n/a	n/a	33,851–34,150*	200–205

Source: Gil *et al.*, 2013: 17 and authors' calculation.

Note
* weighted average.

Table 3.4 Necessary investment for green job creation in selected EU member states

Country	*€ invested/job*	*New jobs created over five years, with 2 per cent investment of GDP*				
		Total	*Energy*	*Construction*	*Transport*	*Manufacturing*
Germany	32,438	4,389,960	377,430	1,870,345	2,147,845	444,340
Bulgaria	8,486	177,330	7,720	60,985	79,370	29,255
Spain	31,400	1,785,340	134,715	505,725	927,945	216,955

Source: authors' calculations and Millennium Institute, 2012.

job creation in Spain is comparable to the cost in Germany, the differences in investment potential (estimated on the basis of GDP) show that Germany is likely to create 2.5 times more green jobs, if the same policies are applied (an investment of 2 per cent of GDP over a five-year period).

It is estimated that a 20 per cent reduction in resource and energy use in *Germany* could lead to an increase in resource productivity of 2.9 per cent per year. The economic benefits would include the creation of more than one million jobs, improved growth rate and an increase of GDP by 12 per cent, with a payback period of one year in the case of materials and six years for energy (GHK, 2009). The variation in levels of resource efficiency between member states suggests that there is scope to improve resource productivity considerably. Increasing resource productivity will provide new jobs as demand grows for techniques that boost productivity or which ease the transition to a low carbon economy. It is hard to measure this impact, as these jobs will be spread throughout the economy, but some lead markets, such as sustainable construction, can be clearly seen as a source of growth.

The *United Kingdom* is an advanced economy with trade focused in services, and a declining manufacturing base over recent decades. It was a large producer of oil and gas once, but declining reserves have increased import dependency. Over the last ten years, three key energy policy issues have emerged: energy security, affordability, and system decarbonisation.[4]

A report by the Royal Commission on Environmental Pollution already in 2000 proposed a voluntary 60 per cent reduction in CO_2 emissions by 2050, recognising the serious challenge of climate change. Over the decade that followed, the government undertook a number of strategic analyses, which further assessed the techno-economic implications of green growth. Based on these strategies the United Kingdom defined a legally binding emission reduction target in 2008.[5]

The level of UK GHG emissions in 2010 was 602 $MtCO_2e$ (excluding international aviation and shipping), 82 per cent of which were CO_2 emissions related to fuel combustion. The largest sources of emissions are power generation, transport and buildings, accounting for 77 per cent of total CO_2 emissions.

Since 1990, GHG emissions have been falling and in 2010 were 22 per cent below 1990 levels. Over half of this reduction (56 per cent) can be attributed to CO_2 emissions, with the remainder from non-CO_2 emissions. Key drivers of the reduction in CO_2 emissions have been the so-called 'dash for gas' and economic restructuring.

Dash for gas is the large-scale take-up of gas for power generation instead of previously dominant coal. Economic restructuring has led to large reductions in emissions from industrial energy use (including in the iron and steel sector) and a general shift to a lower energy-intensive economy.

The government forecasts that the UK economy will continue to grow at around 2.2 per cent annually in the near term and 2.5 per cent between 2022 and 2050. The population is projected to grow from 63.7 million (in 2012) to 70.8 million by 2030 and 76.6 million by 2050. Economy and population growth make emission reductions challenging.

Green growth requires:

- Improving energy efficiency (decreasing the energy intensity of GDP from 2.7 MJ/US$ in 2010 to 0.8 MJ/US$, i.e. 71 per cent reduction).
- Decarbonisation of the power first (by 2013), in order to enable low carbon electricity to become an enabling route for emission reductions in end-use sectors over the 2030–2050 period (decreasing electricity emissions intensity from 503 gCO_2/kWh to 10 gCO_2/kWh in 2050).
- Electrification of end uses (increasing share of electricity in total final energy from current 19 per cent to 47 per cent by 2050).

Greening growth in the United Kingdom relies on a large-scale shift to low carbon power generation (including nuclear, CCS and renewables, particularly wind) and investments of about GBP200 billion to over GBP300 billion. This means an average investment requirement of GBP6.1 billion per year (3.4 GW per year of new capacity) by 2020, increasing to GBP12.3 billion per year (5.7 GW) by 2030 due to the increased construction of capital-intensive low carbon plant, and greater levels of plant retirement.

The mitigation of investment risk is thus critical. The UK government seeks to incentivise large-scale investment in less mature low carbon generation by the electricity market reforms (EMR).

Other risks relate to the viability of CCS, the acceptability of new nuclear facilities and the security of supply related to increasing the share of renewables.

CCS has not yet been sufficiently demonstrated at scale. Demonstrating cost-effective deployment of the first new nuclear reactor in the United Kingdom in 20 years (Hinkley Point C) will be critical to increase acceptability on cost grounds, as well as broader public acceptability. The increasing share of renewables, particularly wind, is accompanied by technical operational challenges including those of demand response, storage and interconnection. However, this requires high infrastructure costs.

The transition to a low carbon economy is challenging and requires a substantial diffusion of a wide range of mitigation options. It is also an opportunity for significant investment in R&D and infrastructure. The emergence of new industries and investment (including investment related to replacing and developing new energy infrastructures) could have major benefits for economic growth. A move towards a lower-carbon technology and resource base could also strengthen energy security. The government has a strong role to play in creating the right investment environment for the transition. Key actions include:

- Maintaining the political will to enact the independently set five-yearly carbon budgets with a corresponding set of implementation policies that are backed across government.
- Through ERM, incentivising investment in low carbon generation that is capital intensive, less mature, possibly more intermittent, and requires significant payback periods.

- Focusing R&D on technologies that the United Kingdom can develop economic benefits from, including offshore renewables but also in automotive industries and other energy technology manufacturers.

Greening the economy could lead to significant improvements. UK businesses could save GBP23 billion a year by using resources more efficiently (Hollins, 2011). The up-front cost of these savings is not high and some involve no cost at all. The investments involved have a pay-back period of one year or less.

The EGS sector in the United Kingdom could create about 50.000 jobs a year, provided that the GIB delivers growth in low carbon economy (Leaton and Reed, 2013: 11).

The TUC estimate shows that meeting 2020 renewables targets would lead to a tenfold increase in jobs in the renewable sector: from around 16,000 positions now to 133,000 jobs in order to design, manufacture, install and operate these new technologies. The wind sector alone (if capacity increases to 30.8 GW) would create 36,000 direct new jobs.

The government's forecasts are more optimistic. It estimates investment opportunities at GBP100 billion and the creation of up to half a million jobs by 2020. However, there are significant differences (tenfold) in data on existing employment and size of the sector. According to the government, the wind power manufacturing sector was worth GBP11.5 billion in 2007/2008 and employed almost 87,500 people. On the other hand, the British Wind Energy Association, estimated the value of the wind industry to be roughly a tenth of this estimate, employing just 5,000.

The United States of America

Green growth based on clean energy, low carbon measures and R&D

The United States is the world's second largest emitter of GHGs, and one of the highest per capita consumers and producers of energy and fossil fuels. Green growth requires profound transformation of the energy sector (the way energy is produced, delivered and used), significant emissions reduction, and a technological transformation of the US infrastructure over time.

The US GHG emissions are dominated by CO_2 from fossil fuel combustion. In 2012, energy-related emissions of all kinds (including fugitive emissions from fuels) accounted for 5,499 $MtCO_2e$, nearly 85 per cent of total gross GHG emissions of 6,526 $MtCO_2e$. Of these, 5,072 Mt (78 per cent) were fossil fuel combustion CO_2. For the United States to do its share in reaching the 2°C target by 2050,[6] the per capita emissions level will need to decrease by an order of magnitude, which is technically feasible (Energy and Environmental Economics Inc., 2014).

The green growth requires:

1 Highly efficient end use of energy in buildings, transportation and industry;
2 Decarbonisation of electricity and other fuels; and

3 Fuel switching of end uses from high-carbon to low carbon supplies.

The challenges and enabling conditions for such growth lie primarily in the realms of cost, policy, public support and resource limitations (primarily biomass availability and CO_2 storage capacity).

Cost reductions for many low carbon measures are often a function of market transformation and high volume production. Continued R&D is also important in many areas, particularly in:

- electrochemistry and nanotechnology (to develop the chemistries, catalysts, and physical matrices fundamental to improvements in batteries, fuel cells, chemical processes, and CO_2 capture); and
- biotechnology and genomics (to advance cellulosic and algal biofuels, biomass synthetic natural gas (SNG) production and biological hydrogen production).

Public support is of key importance for implementation, especially with regard to the siting of low carbon infrastructure.

In the short term, significant improvements in end-use energy efficiency – buildings, appliances, equipment and vehicles – could launch green growth, provided that market failures that prevent the implementation of cost-effective measures are dealt with (e.g. by improvements of codes and standards).

Given the long lifetimes of generation assets, greening the growth requires that all new coal generation includes CCS. For the development of low carbon fuel solutions for industry and freight transport additional R&D is needed.

The efforts in greening the economy resulted in the largest emission cut of any country since 2006, and the energy intensity was reduced by over 15 per cent between 2000 and 2010.

Investments in clean energy in 2011 reached US$45.5 billion, which is a 46 per cent increase on the previous year and the largest of any single country in the world. Almost half (49 per cent) of all new energy capacity built in 2012 was renewable.

The investments were also supported by the US stimulus plan, which allocated US$94 billion to green technologies. Considering that the cost of job creation in the energy sector is between €130,000 and €190,000, this supported roughly a million jobs.

In the United States, aggregated impacts of green investment amounting to 2 per cent of GDP (i.e. US$277 billion) is estimated in the creation of between three and four million jobs, which corresponds to 2.1–2.8 per cent of employment. On average about 18 jobs are created per million US dollars invested.

For the creation of a job, an investment of about €42,896 is necessary, which is about 30 per cent more than in the EU[7] (see Table 3.5).

Table 3.5 Sector specific potential for green job creation in the United States and necessary investment

Sector	*Investment share*	*Annual job creation*	*New jobs created over 5 years*	*Jobs/million USD*	*€ invested/job*
Energy	30	288,386–389,695	1,948,475	4–6	130,833–196,250
Construction	30	1,343,763–1,818,032	9,090,160	20–27	29,074–39,250
Transport	20	1,182,502–1,599,865	7,999,325	26–35	22,429–30,192
Manufacturing	20	250,991–339,576	1,697,888	6–7	112,142–130,833

Source: Millennium Institute, 2012 and authors' calculation.

Canada

Electricity, buildings and transport sector to be decarbonised by 2050

The Canadian context presents a number of challenges related to achieving green growth:

- vast land area drives substantial transportation demand;
- climate drives winter heating and summer cooling demand; and
- the importance of the resource extraction sector to the economy.

To contribute to a path that limits the global increase in temperature to less than 2°C, Canada would need to dramatically reduce CO_2 emissions – by nearly 90 per cent from 2010 by 2050: emissions would need to be reduced from 20.61 tonnes of carbon dioxide equivalent per capita (tCO_2e/cap) in 2010 to less than 2 tCO_2e/cap in 2050, while maintaining strong economic growth (Bataille *et al.*, 2014).[8]

This would require major shifts in technology adoption, energy use and economic structure.

Green growth would enable that the electricity, transportation and building sectors are decarbonised by 2050. This requires:

- reinforced and deepened energy efficiency improvement trends in all energy end-uses;
- eventual decarbonisation of the electricity sector;
- fuel switching to lower carbon fuels and decarbonised energy carriers (e.g. electricity, transport biofuels and hydrogen); and
- direct GHG reduction for industrial processes and thermal heat generation (e.g. via CCS and process changes).

The fossil fuel production and mineral extraction sectors in Canada are export oriented and play a major role in the Canadian economy. The reduction of emissions in these sectors depends primarily on available technologies (such as CCS technologies), i.e. greening the Canadian growth heavily depends on technological innovation and deployment, which requires investments, innovation and global research cooperation, as well as public acceptance.

Mexico

Green growth should be driven by services and smart urban development

The use of fossil fuels has been increasing in Mexico, which leads to rising GHG emissions. The population is slowly stabilising and is projected to be 151 million by 2050 (Buira and Tovilla, 2014).

In this context, green growth enables continued economic growth, the mitigation of climate change and related co-benefits (local health improvement, economic savings, greater productivity, poverty reduction and social inclusion).

Green growth should be driven by the services sector, which could account for nearly 70 per cent of national GDP by 2050 (in 2010 this sector represented around 62 per cent of the total GDP). The services sector is less energy and CO_2 intensive than other economic activities in Mexico. Thus, a shift in economic structure should decrease GHG emissions.

There is a strong correlation between the increased GDP per capita and the growth of medium-sized cities. Urban population reached 72 per cent in 2010, and is expected to be close to 83 per cent by 2030. Smart urban development is thus a key to a transition towards more efficient and sustainable green growth schemes in Mexico.

Such transition requires changes of current policies – primarily related to energy subsidies and the introduction of carbon taxes to imports and exports of fossil fuels. Resources to fund the transition, and the technological availability of cost-effective options (e.g. for CCS, electric vehicles, solar power harvesting, biofuels production, the development of zero carbon or carbon negative agriculture and forestry techniques to support the production of sustainable bioenergy crops and reduce emissions from these sectors) are also necessary. Also, the adoption of better practices in urbanisation and territorial planning is necessary, since better organised cities could induce the behavioural changes needed.

Modelling the effects of greening the economy in Mexico shows that the expected results are increased employment and reduced output, e.g. the effect of substituting 10 per cent of final demand in agriculture, electric energy, construction, manufacturing industry and tourism with corresponding green activities was estimated as a net decrease in output of MXN 8,653 million (which represents 0.04 per cent of production in Mexico) and a net increase of 736,376 jobs, representing a 2 per cent increase in total employment.

Most green activities are more labour intensive, and thus have a higher employment multiplier than their conventional counterpart. The sectors that have a higher employment multiplier than their conventional counterparts are organic agriculture, green construction, renewable energy, manufacturing industry, waste management, and temporal lodging services.

For example, for each million-peso increase in agricultural output, the agriculture activity generates 16 jobs, while organic agriculture generates 45 jobs through the same increase in final demand. Thus, the green activities analysed have an employment multiplier higher or equal to their conventional activities. Thus, 1,815 million direct environmental jobs create another 971 million indirect jobs, adding to a total of 2.786 million jobs. In addition to the employment effects, organic agriculture is an inclusive rural development mechanism. Work conditions in organic agriculture are better than in traditional agriculture due to the elimination of harmful chemicals. In order to promote organic agricultural activities, sectorial policies and incentives are needed to stimulate domestic demand (which is low) and organic agriculture, which currently depends on international markets. The achievement of such growth requires an increase in organic training programmes for farmers.

Considering the national climate change policy and its objectives for 2030, renewable energy generation is the sector with the most significant potential for achieving environmental and economic goals, since it is the economy's second most productive sector. Furthermore, the growth of renewable energy production is mainly based on services, construction and manufactured inputs, strengthening local economic activities. Indirect jobs generated by this sector double the amount of direct jobs and employment in these activities can be considered as green jobs, since the Decent Work Index equals 79, indicating good working conditions.

Turkey

Green growth as a part of EU accession process

Between 2001 and 2010 real GDP increased by more than 50 per cent.[9] The average growth rate was nearly 7 per cent during the period 2003–2007 (about 4 per cent during the 1990s). Growth resumed rapidly after the 2008–2009 global crisis, at 9.2 per cent in 2010 and 8.5 per cent in 2011. Per capita income reached US$10,444, and inflation came down to under 10 per cent (from around 70 per cent). Economic growth is still coupled with energy consumption.[10]

Turkey's economy is increasingly dominated by the services sector, which contributes to about 68 per cent of the GDP, followed by industry (23 per cent) and agriculture (9 per cent). Further movement towards the expansion of services at the expense of industry and agriculture is expected to result in declines in intensely-used resource (including land, water and energy) and increased emissions (pollution and waste).

Turkey has prepared a 'vision paper' on green growth as its contribution to the Rio+20 Conference. The vision focuses on:

1. showcasing its efforts to achieve rapid economic growth, while integrating environmental sustainability requirements in key sector policies and programmes; and
2. as part of the preparation of its tenth National Development Plan (NDP) (2014–2018) and beyond, identifying potential additional policy and institutional measures to further green its economy without eroding its growth potential in the short-term.

In addition, Turkish development goals until 2023 include:

- to become one of the world's ten largest economies;
- to reach GDP of US$2 trillion;
- to exceed the foreign trade volume US$1 trillion;
- to reach per capita income US$25,000;

- to reduce unemployment to 5 per cent;
- to complete full membership negotiations with the EU;
- to further develop Istanbul as a leading international financial hub; and
- to become the leading manufacturing and service provider both in the region and beyond.

Alignment with the EU environmental acquis, as part of EU accession process, requires greening its economy. Environmental compliance is costly and requires investments covering water and air pollution, waste and chemicals management, biotechnology, radiation protection and nature conservation. The EU Integrated Environmental Approximation Strategy (2007–2023) estimated the total cost to be about €59 billion – of which nearly 58 per cent will be in the water sector. Currently, Turkey is spending below 0.5 per cent of GDP on the environment. During a few peak investment years this share is expected to reach 2 per cent of GDP.

There are six industrial manufacturing sectors – automotive, construction, electronics, iron and steel, machine industry and white goods, and agriculture – that are considered strategic for green growth.[11]

The Turkish automotive industry is one of the economy's pioneering and most dynamic sectors, with an average annual growth rate of 4.2 per cent from 2000 to 2010. The automotive industry is a major contributor to air pollution (e.g. volatile organic compounds (VOC) and CO_2 emissions), solid waste (e.g. packaging, metal scrap, parts and consumables), hazardous waste (e.g. paint sludge, oils, batteries, electronics), and waste water (e.g. oils and chemical discharges, especially from uncontrolled/unregistered repair shops).

The greening programme of the industry includes:

1. energy efficiency (e.g. insulation, co-generation, and reuse of heat from boilers, and gradually switching to renewable energy);
2. adoption of improved vehicle and motor technologies (e.g. lighter body material, tyre performance, direct fuel injection, multi-fuel and hybrid fuel vehicles, hybrid cars, EU vehicle emissions standards); and
3. improved waste management (including separation and recycling) and physical–chemical water treatment plants in all production facilities.

There are also measures to reduce the environmental impact of the service life of vehicles (waste management during maintenance and repair, and end-of-life scrap recycling), and more specifically towards compliance with the EU End-of-Life Vehicle Directive (ELV 2000/53/EC) which requires EU member states to ensure that a minimum of 85 per cent of vehicles are reused or recovered (including energy recovery) by 2015.

The iron and steel industry is the largest industrial emitter of CO_2 in the world. With some 30 million tons in 2010, Turkey ranks tenth in the world and second in Europe in terms of total steel output. Turkish production has almost doubled since 2001 (second fastest increase behind China). Turkey's steel production is relatively green, because 80 per cent of its total production capacity (42.7 million tons in 2010) uses electric arc furnaces (EAF) technology, while the rest is in dirtier, less energy-efficient integrated steel plants (ISP) technology.[12]

The industry has the potential for further greening through reducing carbon emissions in the production process (especially ISP), and reducing the intensity of other pollutants (e.g. heavy metals).[13]

The greening potential of the construction sector can be illustrated by the cement industry. Both globally and in Turkey, the cement industry is a major emitter of carbon (5 per cent of total GHG emissions). Turkey is among the top ten cement manufacturing countries and a leading world cement exporter. Key pollution issues in the sector with direct health impacts include high levels of chromium (a heavy metal) with concentrations 15 times the level called for by the EU REACH Directive, and high levels of CO_2, NO_x (mono-nitrogen oxides), dust, and sulphur emissions.

While the cement sector has undergone significant modernisation, there remains potential for additional greening, including in terms of resource use and efficiency and pollution reduction.

The machine industry contributes to nearly 4.5 per cent of GDP, approximately 200,000 jobs and its total production is about US$25 billion. Because of its wide span (agriculture, electrical, food, construction mining and textile machinery), the sector's multiplier effect is important for encouraging investment and demand for intermediate goods and services. It also has considerable impact on national competitiveness as a whole. Since 1990, the industry has consistently grown about 20 per cent per annum. The potential for greening is yet to be defined, primarily by energy audits to identify potential energy efficiency and/or fuel substitution measures, as well as measures to reduce and recycle waste materials (especially metals and plastics), waste water, solid waste and hazardous substances.

Turkey's household appliances industry, which started as an assembly industry with foreign partners, has achieved tremendous progress and is now a flagship of the Turkish economy. The industry currently contributes about 1 per cent of GDP and US$2.7 billion exports in 2010. Between 2000 and 2010, the sector increased its exports sevenfold. The sector has good potential for further greening because of its link to market demand and an emphasis on innovation and R&D. In addition to energy efficiency in production, the increased collection and the reuse of metals and plastics as major components in the sector would have a positive impact on reducing costs and decreasing pollution.

Beginning with foreign license agreements with companies mostly from the United States or Europe, Turkey's electronics industry has changed and matured considerably since the late 1980s. The sector has been adding value to the Turkish economy as a flagship of innovation, quality and competitiveness and export-driven growth (1.5 per cent contribution to GDP and US$6 billion exports in 2010). From 2000–2010, the sector increased its exports nearly fourfold. Hazardous waste (including heavy metals such as lead, cadmium and mercury) and energy consumption are the most important issues in the manufacturing of consumer electronics.

Much of the old equipment and waste is collected by informal scrap dealers and sold on secondary markets, which is an opportunity for greening this sector.

Agriculture has been growing slightly slower than the population, at an average rate of around 1.2–1.5 per cent per year during recent decades. The sector is the largest employer in the country (84 per cent of women living in rural areas were employed in the sector in 2009). Population growth (1.35 per cent per year since 2000) and increasing per capita consumption are the main drivers of demand. Increasing wealth is expected to boost demand for products with a higher resource footprint, such as meat. Thus the resource base may limit agricultural potential. Greening the agriculture will require improving the efficiency of resource use and resilience to climate risk.

Brazil

Green growth requires specific measures related to forestry

Population and economic growth have been consistent drivers of increased energy-related CO_2 emissions, whereas the energy-related CO_2 intensity per unit of GDP increased from 1990 to 2000 but decreased from 2000 to 2010.

Brazil uses a lot of clear energy sources, and thus has relatively low per capita energy-related GHG emissions. Major emissions have been historically concentrated in agriculture, forestry, and other land use, related mostly to deforestation, crop growing and livestock. The Brazilian population has increased from 145 to 191 million people between 1990 and 2010.

GHG emissions increased from 1.4 billion metric tons CO_2 equivalent ($GtCO_2e$) in 1990 to 2.5 $GtCO_2e$ in 2004, followed by a substantial reduction (by half), reaching 1.25 $GtCO_2e$ in 2010, thanks to the sharp fall of deforestation. As a result, the share of CO_2 in the GHG emissions mix declined from 73 to 57 per cent between 2005 and 2010.

Greening the Brazilian economy by 2030 requires decarbonisation outside the energy sector. In addition to standard measures (in energy, the transport sector, industry and buildings) green growth in Brazil requires the implementation of specific measures related to forestry, primarily the afforestation programs.

It is estimated that Brazil could cut greenhouse emissions by 37 per cent without compromising growth, jobs and development goals. Also, a new land use dynamic in Brazil would reduce deforestation by up to 68 per cent by 2030 (De Gouvello, 2010)

Green growth requires transformation and structural changes that have high upfront costs. A 'low carbon scenario' would require additional investments of around US$400 billion over 20 years (World Bank, 2010). Thus, a strong public awareness of the potential dangers of climate change and the pitfalls of inaction is necessary. Given the low savings capacity of the Brazilian economy, the financial flow will need to largely come from outside of Brazil (Lebre La Rovere and Gesteira, 2014).

Final remarks

The green growth theory adds the issue of natural capital into traditional economic growth theory. The natural capital refers to the ecosystems that sustain the economy and society. With the emergence of environmental economics in the 1970s, natural capital was increasingly integrated into traditional economic growth models. But it was done so largely on a 'negative' basis, in the sense that natural capital – or rather the boundaries imposed by limited natural capital – imposed potential constraints on economic growth.

In contrast to the traditional (Korean) approach that considers green growth as the most advanced type of economic growth, there is also an (European) approach that considers green growth as a consequence of the creation of green jobs and as the way to achieve a green economy.

Despite the fact that there is no consensus on cause and consequence, there is common understanding that green growth is expected to create economic and environmental benefits and thus contribute to sustainable development.

Notes

1 IDDRI is a non-profit policy research institute based in Paris. It is an independent institute that mobilises resources and expertise to disseminate relevant scientific ideas and research ahead of negotiations and decision making processes.
2 ESCAP considers green low carbon growth as a single concept – a new economic development path (ESCAP, 2012).
3 The incident occurred when the Fukushima nuclear power plant was hit by a tsunami triggered by an earthquake and that resulted in the meltdown of three of the plant's six nuclear reactors.
4 This section is based on Pye and Anandarajah, 2014.
5 Climate Change Act 2008, specifies that a GHG reduction target of 80 per cent is to be achieved by 2050 (relative to 1990 levels). A set of five-year carbon budgets is independently proposed and monitored by the Committee on Climate Change.
6 Limiting the rise in average global surface temperature to 2°C (3.6°F) was for more than a decade seemingly a simple target in international climate-policy discussions. The two-degree target first gained prominence in the early 1990s, when a number of international scientific panels suggested the limit as a way to maintain the relatively stable climate conditions that humans (and other species) had adapted to over the previous 12,000 years and to prevent some of the worst impacts of climate change-driven drought, heat waves and rising sea levels. The 2009 Copenhagen Accord (the result of that year's UN Climate Change Conference) enshrined a two-degree rise in the global average temperature as the threshold of 'dangerous' human interference in the climate system.
7 For conversion, the average annual exchange rate between dollars and euros in 2010 is used (0.785), based on IRS data (online, available at: www.irs.gov/Individuals/International-Taxpayers/Yearly-Average-Currency-Exchange-Rates, accessed 10 November 2014).
8 Increase of GDP per capita from US$27,288 in 2010 to US$78,882 in 2050, while population is expected to grow from 33.8 million to 48.3 million.
9 This section is based on Bouzaher *et al.* (2013).
10 It is among 25 countries that display the fastest rate of growth in the industrial use of energy sources. The consumption of electricity per capita in Turkey has increased sixfold from 1980 to 2005, and from 300 kWh in 2005 to 400 kWh per person in 2010.

11 The strategic importance of the manufacturing sectors stems from the following: (a) they are highlighted as key drivers of growth, (b) each sector is subject to compliance with key applicable EU Environmental *Acquis* Directives, (c) they have a large potential for 'greening' in terms of increasing input efficiency and lowering pollution intensity, and each sector has a strategy from which to draw information for analytical purposes. Agriculture, as a non-manufacturing sector, is added owing to its important role in the economy and its potential for 'greening'.

12 There are two widely used production technologies: ISPs comprised of blast furnaces and basic oxygen furnaces, and EAFs. While ISP processes use coke to reduce iron ore in blast furnaces, the EAF process primarily uses scrap metal (melted through high electric voltage).

13 There is also the potential to apply innovative solutions for greening in the iron sector, e.g. Dubai uses the heat from its steel industry for sea water desalination, promoting a circular economy.

References

Allen, C. and Clouth, S., 2012. 'A Guidebook to the Green Economy'. Issue 1: *Green Economy, Green Growth, and Low-carbon Development – History, Definitions and a Guide to Recent Publications*. Division for Sustainable Development, New York: UNDESA.

Bouzaher, A., Guadagni, M., Sahin, S., Meisner, C., Raiser, M. and Kasek, L.P., 2013. *Turkey Green Growth Policy Paper: Towards a Greener Economy*. Washington, DC: World Bank.

Bataille, C., Chan, J., Sawyer, D. and Adamson, R., 2014. *Pathways to Deep Decarbonisation, Chapter Canada*. 2014 report SDSN and IDDRI. Online, available at: http://unsdsn.org/wp-content/uploads/2014/09/DDPP_2014_report_Canada_chapter.pdf (accessed 15 October 2014).

Buira, D. and Tovilla, J., 2014. *Pathways to Deep Decarbonisation, Chapter Mexico*. 2014 SDSN and IDDRI report. Online, available at: http://unsdsn.org/wp-content/uploads/2014/09/DDPP_2014_report_Mexico_chapter.pdf (accessed 15 October 2014).

Canzi, G. and Tkacik, J., 2014. 'Environment and Economic Growth: Inevitable Conflict?' In *Global Development Goals: Partnerships for Progress*. London: United Nations Association UK. Online, available at: http://17aa47148cdcdf8b5c51-da5ed-784d101708d617ec977f6449487.r27.cf2.rackcdn.com/Global%20Development%20Goals%202014.pdf accessed (2 November 2014).

Capozza, I., 2011. 'Greening Growth in Japan'. *OECD Environment Working Papers, No. 28*. Paris: OECD Publishing. Online, available at: http://dx.doi.org/10.1787/5kggc0rpw55l-en (accessed 14 October 2014).

De Gouvello, C., 2010. *Brazil Low-carbon Country Case Study*. Washington, DC: World Bank Group. Online, available at: http://siteresources.worldbank.org/brazilextn/Resources/Brazil_LowcarbonStudy.pdf (accessed 4 October 2014).

EC, 2012. *Exploiting the Employment Potential of Green Growth*. SDW (2012) 92. 18.4.2012. Brussels.

EC, 2014. *Green Action Plan for SMEs, enabling SMEs to turn Environmental Challenges into Business Opportunities*. COM (2014) 440 final. Brussels, 2.7.2014.

Energy and Environmental Economics, Inc., 2014. *Pathways to Deep Decarbonisation, Chapter United States*. 2014 report SDSN and IDDRI. Online, available at: http://unsdsn.org/wp-content/uploads/2014/09/DDPP_2014_report_United_States_chapter.pdf (accessed 15 October 2014).

ESCAP, 2012. *Low Carbon Green Growth Roadmap for Asia and the Pacific, Turning Resource Constraints and the Climate Crisis into Economic Growth Opportunities*. Bangkok: UN. Online, available at: www.timhaahs.com/wp-content/uploads/2014/03/Low-Carbon-Green-Growth-Roadmap-for-Asia-and-the-Pacific.pdf (accessed 17 October 2014).

European Parliament and the Council, 2013. *Decision No 1386/2013/EU on a General Union Environment Action Programme to 2020 'Living well, within the Limits of our Planet'*. Online, available at: http://faolex.fao.org/docs/pdf/eur129696.pdf (accessed 1 November 2014).

Fei, T., Qiang, L., Alun, G., Xi, C., Chuan, T. and Xiaoqi, Z., 2014. *Pathways to Deep Decarbonisation, Chapter China*. 2014 report SDSN and IDDRI. Online, available at: http://unsdsn.org/wp-content/uploads/2014/09/DDPP_2014_report_China_chapter.pdf (accessed 15 October 2014).

GGGI (Global Green Growth Institute), 2012. *Agreement on the Establishment of the Global Green Growth Institute*. Online, available at: http://gggi.org/wp-content/uploads/2012/10/Agreement-on-the-Establishment-of-the-GGGI.pdf accessed (29 September 2014).

GHK, 2009. *The Economic Benefits of Environmental Policy Global Climate Network, Low Carbon Jobs in an Interconnected World.* Online, available at: www.ippr.org/publications/low-carbon-jobs-in-an-interconnected-world (accessed 2 December 2014).

Hall, M., 2014. *EU Tables 'Circular Economy' Package with Zero-landfill Goal*. Euractiv 03/07/2014. Online, available at: www.euractiv.com/sections/sustainable-dev/eu-tables-circular-economy-package-zero-landifll-goal-303259 (accessed 1 November 2014).

Hollins, O., 2011. *Resource Efficiency*. Research project. UK Department for Environment, Food and Rural Affairs. Online, available at: www.gov.uk/government/news/research-shows-companies-can-save-money-by-helping-the-environment (accessed 2 April 2015).

HM Government, 2011. *Skills for a Green Economy, a Report on the Evidence*. London, UK. Online, available at: www.gov.uk/government/uploads/system/uploads/attachment_data/file/32373/11-1315-skills-for-a-green-economy.pdf (accessed 2 November 2014).

Hyon-Hee, S., 2013. 'Korea Eyes Era of "Green Growth 2.0"'. *Korea Herald*. 11 October. Online, available at: www.koreaherald.com/view.php?ud=20131110000342 (accessed 2 October 2014).

ILO, 2014a. *World of Work 2014*. Report. Developing with Jobs. Geneva. Online, available at: http://ilo.org/global/research/global-reports/world-of-work/2014/WCMS_243961/lang--en/index.htm (accessed 26 March 2015).

ILO, 2014b. R*eport III – Report of the Conference – 19th International Conference of Labour Statisticians*, Geneva, 2–11 October 2013. Online, available at: www.ilo.org/stat/Publications/WCMS_234124/lang--en/index.htm (accessed 26 March 2015).

ILO 2014c. *Frequently Asked Questions on Green Jobs*. Online, available at: www.ilo.org/global/topics/green-jobs/WCMS_214247_EN/lang--en/index.htm (accessed 26 March 2015).

IMF, 2010. *IMF Proposes 'Green Fund' for Climate Change Financing*. Online, available at: www.imf.org/external/pubs/ft/survey/so/2010/NEW013010A.htm (accessed 2 November 2014).

IRS, 2014. *Yearly Average Currency Exchange Rates. Translating Foreign Currency into US Dollars*. Online, available at: www.irs.gov/Individuals/International-Taxpayers/Yearly-Average-Currency-Exchange-Rates (accessed 19 November 2014).

ITUC, 2014a. *Official Website*. Online, available at: www.ituc-csi.org/ (accessed 29 November 2014).

ITUC, 2014b. *Economic Briefing*. Online, available at: www.ituc-csi.org/IMG/pdf/ituc_economicoutlook_n3_en_4_.pdf (accessed 29 November 2014).

Jaeger, C.C., Paroussos, L., Mangalagiu, D., Kupers, R., Mandel, A. and Tabara, J.D., 2011. *A New Growth Path for Europe. Generating Prosperity and Jobs in the Low-carbon Economy*. Synthesis Report. A study commissioned by the German Federal Ministry for the Environment, Nature Conservation and Nuclear Safety. Online, available at: www.pik-potsdam.de/members/cjaeger/a_new_growth_path_for_europe__synthesis_report.pdf (accessed 29 November 2014).

Jones V., 2009. *The Green-collar Economy*. New York: HarperCollins.

Jotzo, F., Skarbek, A., Denis, A., Jones, A., Kelly, R., Ferraro, S., Kautto, N., Graham, P., Hatfield-Dodds, S. and Adams, P., 2014. *Pathways to Deep Decarbonisation. Australia Chapter*. 2014 report SDSN and IDDRI. Online, available at: http://unsdsn.org/wp-content/uploads/2014/09/DDPP_2014_report_Australia_chapter.pdf (accessed 15 October 2014).

Kainuma, M., Oshiro, K., Hibino, G. and Masui, T., 2014. *Pathways to Deep Decarbonisation, Japan Chapter*. 2014 report SDSN and IDDRI. Online, available at: http://unsdsn.org/wp-content/uploads/2014/09/DDPP_2014_report_Japan_chapter.pdf (accessed 15 October 2014).

Kashara, S., 2013. *The Asian Developmental State and the Flying Geese Paradigm*. UNCTAD Discussion paper No. 213, November 2013.

Krugman, P., 2010. 'Building a Green Economy'. *New York Times* 4 July. Online, available at: www.nytimes.com/2010/04/11/magazine/11Economy-t.html?ref=magazine.

Leaton, J. and Reed, H., 2013. *A Bank for the Future*. London: WDM. Online, available at: www.banktrack.org/manage/ems_files/download/a_bank_for_the_future_maximizing_public_investment_in_a_low_carbon_economy/100810_a_bank_for_the_future.pdf.

Lebre La Rovere, E. and Gesteira, C., 2014. *Pathways to Deep Decarbonisation, Chapter Brazil*. 2014 report SDSN and IDDRI. Online, available at: http://unsdsn.org/wp-content/uploads/2014/09/DDPP_2014_report_Brazil_chapter.pdf (accessed 15 October 2014).

Millennium Institute, 2012a. *Growing Green and Decent Jobs*. ITUC. Online, available at: www.ituc-csi.org/IMG/pdf/ituc_green_jobs_summary_en_final.pdf (accessed 19 November 2014).

Millennium Institute, 2012b. *United States Country Profile*. Online, available at www.ituc-csi.org/IMG/pdf/us_country_profile_mar2012.pdf (accessed 14 October 2014).

OECD, 2009. *Declaration on Green Growth*. C/MIN(2009)5/ADD1/FINAL 25 June 2009.

OECD, 2010. *Interim Report of the Green Growth Strategy: Implementing our Commitment for a Sustainable Future*. Online, available at: www.oecd.org/greengrowth/45312720.pdf (accessed 25 September 2014).

OECD, 2011. *Towards Green Growth*. Online, available at: www.oecd.org/greengrowth/48224539.pdf (accessed 24 September 2014).

Piketty, T., 2014. *Capital in the Twenty First Century*. Cambridge, MA: Harvard University Press.

Pye, S. and Anandarajah, G., 2014. *Pathways to Deep Decarbonisation, Chapter United Kingdom*. 2014 report SDSN and IDDRI. Online, available at: http://unsdsn.org/wp-content/uploads/2014/09/DDPP_2014_report_United_Kingdom_chapter.pdf (accessed 15 October 2014).

Sachs, J. and Tubiana, L., 2014. *Pathways to Deep Decarbonisation.* 2014 report SDSN and IDDRI. Online, available at: http://unsdsn.org/wp-content/uploads/2014/09/DDPP_Digit.pdf (accessed 15 October 2014).

UAE Ministry of the Environment and Water, 2013. *UAE to Host First Global Conference on Partnership for Acton on Green Economy*, Statement issued by the UAE Ministry of the Environment and Water at a press conference on 17 December 2013. Online, available at: www.unep.org/greeneconomy/Portals/88/PAGE/UAE%20press%20release%20-%2017%20dec%202013.pdf (accessed 2 November 2014).

United Nations, 2012. *A Guidebook to Green Economy.* Online, available at: http://sustainabledevelopment.un.org/content/documents/GE%20Guidebook.pdf (accessed 6 November 2014).

World Bank, 2012a. *Inclusive Green Growth: the Pathway to Sustainable Development.* Washington. Online, available at: http://siteresources.worldbank.org/EXTSDNET/Resources/Inclusive_Green_Growth_May_2012.pdf (accessed 13 November 2014).

World Bank, 2012b. *China 2030.* Online, available at: www.worldbank.org/content/dam/Worldbank/document/China-2030-complete.pdf (accessed 21 November 2014).

WEF (World Economic Forum), 2014a. *Global Risks 2014.* Online, available at: www3.weforum.org/docs/WEF_GlobalRisks_Report_2014.pdf (accessed 25 November 2014).

WEF, 2014b. *The World Economic Forum's Outlook on the Global Agenda 2014.* Online, available at: www3.weforum.org/docs/WEF_GAC_GlobalAgendaOutlook_2014.pdf (accessed 25 November 2014).

4 Sustainable development

The concept

The term sustainable development became popular after it was used by the Brundtland Commission in its 1987 report 'Our Common Future', which defined sustainable development as 'development that meets the needs of the present without compromising the ability of future generations to meet their own needs'.

Sustainable development integrates economic, social and environmental objectives. It recognises that economic growth is necessary, but not sufficient for development. The concept of sustainable development was integrated in the Agenda 21 and other key documents produced at the 1992 Rio Summit[1] and further elaborated in other global development documents (see Table 4.1).

The final declaration of the UN Millennium Summit, the Millennium Declaration, adopted in 2000 and signed by 189 countries, reaffirmed the Agenda 21 and outlined the international agenda for development. The Millennium Declaration lists eight interdependent MDGs to be achieved by 2015, including the eradication of poverty and environmental sustainability (see Table 4.2).

Table 4.1 Key initiatives towards global sustainable development

Year	*United Nations sustainable development initiatives*
1972	UN International Conference on Human environment in Stockholm (establishment of UNEP-UN environmental programme)
1987	UN Brundtland Report 'Our Common Future'
1992	UN Rio de Janeiro – operationalisation of the concept of sustainable development (action plans and measures)
	Agenda 21 – environment must be in all policies
2000	UN General Assembly – Millennium Declaration –MDGs
2002	World Summit on SD in Johannesburg (lack of sanctions for the lack of actions)
2002	UN Resolution – the Decade of Education for Sustainable Development
2012	Rio de Janeiro – Rio +20 UN Conference on Sustainable development – 'green jobs', 'The Future We Want'
2013/2014	UNGS – new sustainable development goals – beyond 2015

Source: authors' compilation.

Table 4.2 MDGs and proposals for SDGs

MDGs	*SDGs (proposal)*
1 Eradicate extreme poverty and hunger • Reduce by half the proportion of people whose income is less than USD 1 a day • Achieve full and productive employment and decent work for all, including women and young people • Reduce by half the proportion of people who suffer from hunger 2 Achieve universal primary education • Ensure that all boys and girls complete a full course of primary schooling 3 Promote gender equality and empower women • Eliminate gender disparity in primary and secondary education preferably by 2005, and in all levels of education no later than 2015 4 Reduce child mortality • Reduce by two-thirds the mortality of children under five 5 Improve maternal health • Reduce maternal mortality by three-quarters • Achieve universal access to reproductive health 6 Combat HIV/AIDS, malaria and other diseases • Halt and reverse the spread of HIV/AIDS • Achieve, by 2010, universal access to treatment for HIV/AIDS for all those who need it • Halt and reverse the incidence of malaria and other major diseases 7 Ensure environmental sustainability • Integrate principles of sustainable development into country policies and programmes; reverse the loss of environmental resources • Reduce biodiversity loss, achieving, by 2010, a significant reduction in the rate of loss • Halve the proportion of people without access to safe drinking water and basic sanitation • Improve the lives of at least 100 million slum dwellers by 2020 8 Develop a global partnership for development • Develop further an open, rule-based, predictable, non-discriminatory trading and financial system • Address special needs of the least developed countries, landlocked countries and small island developing states • Deal comprehensively with developing countries' debt • In cooperation with pharmaceutical companies, provide access to affordable essential drugs in developing countries • In cooperation with the private sector, make available the benefits of new technologies, especially information and communications technologies	1 End poverty in all its forms everywhere 2 End hunger, achieve food security and improved nutrition and promote sustainable agriculture 3 Ensure healthy lives and promote well-being for all at all ages 4 Ensure inclusive and equitable quality education and promote lifelong learning opportunities for all 5 Achieve gender equality and empower all women and girls 6 Ensure availability and sustainable management of water and sanitation for all 7 Ensure access to affordable, reliable, sustainable and modern energy for all 8 Promote sustained, inclusive and sustainable economic growth, full and productive employment and decent work for all 9 Build resilient infrastructure, promote inclusive and sustainable industrialization and foster innovation 10 Reduce inequality within and among countries 11 Make cities and human settlements inclusive, safe, resilient and sustainable 12 Ensure sustainable consumption and production patterns 13 Take urgent action to combat climate change and its impacts 14 Conserve and sustainably use the oceans, seas and marine resources for sustainable development 15 Protect, restore and promote sustainable use of terrestrial ecosystems, sustainably manage forests, combat desertification, and halt and reverse land degradation and halt biodiversity loss 16 Promote peaceful and inclusive societies for sustainable development, provide access to justice for all and build active, accountable and inclusive institutions at all levels 17 Strengthen the means of implementation and revitalize the global partnership for sustainable development

Source: authors' compilation, based on UN 2010, and Open Working Group, 2014.

The interdependence of goals, activities and results at local, regional and global level makes formulating, costing, financing and implementing adequate policies difficult. Despite these difficulties, the MDGs articulate a long-term perspective and provide direction for the implementation of development strategies. They call specifically for integration of the principles of sustainable development into country policies and programs.

Based on the Agenda 21 and MDGs, each country was expected to adopt its national sustainable development strategy (NSDS) and to begin their implementation by 2005. The strategic policy frameworks that include sustainable development issues (e.g. generic NSDSs – EU, Australia; sustainable development included in other policy strategies – Japan, etc.) vary significantly among the countries and world regions (see Table 4.3).

The World Summit on Sustainable Development held in Johannesburg in 2002 tried to assess the progress in achieving the MDGs and the Agenda 21. It concluded that both documents provide strong support for the principle of sustainable development, but that implementing instruments and sanctions for inaction are missing.

Key global development documents (the Rio Declaration on Environment and Education adopted in Rio in 1992, the Agenda 21, the Johannesburg Plan of Implementation (JPoI) from 2002 and the MDGs – in particular Goal 2: achieve universal primary education, see Table 4.2) emphasise quality education and information exchange as a major catalyst for sustainable development. Thus, the UN adopted a declaration that proclaimed the ten-year period from 2005 to 2014 the Decade of Education for Sustainable Development. Such education means including key sustainable development issues (such as climate change, disaster risk reduction, biodiversity, poverty reduction, and sustainable consumption) into teaching and learning. Education for Sustainable Development consequently promotes competencies like critical thinking, imagining future scenarios and making decisions in a collaborative way. In cooperation with other international organisations and representatives of the educational sector in individual countries,

Table 4.3 Strategic policy frameworks for sustainable development

Country	*Description*
EU	Sustainable Development Strategy
Canada	Sectoral sustainable development strategies Federal sustainable development strategy
United States	Action plan for sustainable development (not formally adopted)
Japan	National Agenda 21 Action Plan, Basic environmental plan and New growth strategy
Mexico	National Development Plan
Brazil	Brazilian Agenda 21
Turkey	Development Action Plan
China	China's Agenda 21, National Five-Year Plan 2011–2015

Source: author's compilation.

the UN encourages countries to change national educational programmes and integrate sustainable development into existing strategies and action plans for education.

The United Nations Conference on Sustainable Development in Rio de Janeiro in 2012 (so called Rio+20) examined the strengths and weaknesses in the management of global sustainable development policy and set targets for future initiatives and challenges.

The results of the conference are summarised in the document 'The Future We Want'. It emphasises the need for further strengthening of global cooperation and partnership in the promotion of 'green' growth as a fundamental way of combating poverty and a means of establishing a balance between economic, social and environmental dimensions.

One of the main outcomes of the Rio+20 Conference was the agreement to launch a process to develop a set of SDGs, which will build upon the MDGs. The proposal for the new global development goals includes 17 goals (see Table 4.2). It is based on correcting the deficiencies identified in the formulation and progress towards achievement of the MDGs, which has been uneven. The MDGs helped in raising awareness about the need to combat poverty and achieve sustainable development. A number of initiatives and projects at national and local levels of governance were launched. Poverty was reduced (especially in Asia), health improved, as well as access to primary education, but less success has been achieved in strengthening employment policy, the fight against climate change and combating hunger and malnutrition. The interdependence between MDGs is not sufficiently considered, nor are the specific national problems. For instance, the problem of poverty in developed countries has not been covered by any goal. Thus, the main challenge for the period post-2015 will be to strengthen the global partnership, in achieving the ultimate vision of development – the eradication of poverty in the world by 2030. The main goal (to end poverty in all its forms everywhere by 2030) is far more ambitious than in the MDGs, which deal only with extreme poverty and do no tackle poverty in urban areas in developed countries at all. The formulation of this overriding goal also illustrates the most important changes in the formulation of goals for the post-2015 period: the development of targets that are universal and of which the achievement is the common aim of all countries, but which also need to be adapted to the specificities of each country.

The successful implementation of SDGs for the entire world by 2030 would mean reducing the number of poor and extremely poor people by 1.2 billion, increasing the number of workers with good jobs and a good quality of life by 470 million, increasing the number of people who have access to electricity by 1.2 billion, reducing the number of people who are affected by natural disasters by 220 million, and so on.

To achieve such improvements in quality and conditions of life, it is necessary to pay more attention to raising stable, long-term finance for development, define priorities in areas that go beyond aid, and infuse global partnerships and cooperation into all the goals. This is included in Goal 17: strengthen the means of implementation and revitalise the global partnership for sustainable development.

At least three out of 17 proposed SDGs are green, despite different wording:

- Goal 8: promoting sustained, inclusive and sustainable economic growth, full and productive employment and decent work for all promotes green growth and jobs.
- Goal 9: building resilient infrastructure, promoting inclusive and sustainable industrialisation and fostering innovation fosters greening the economy.
- Goal 11: making cities and human settlements inclusive, safe, resilient and sustainable deals with greening the cities.

This indicates that the 'green' growth does not replace the concept of sustainable development; it is one of its components. This is also reflected in the various documents of international organisations, such as the ILO, the World Bank and the OECD.

According to the ILO, green jobs contribute to the creation of a 'double dividend' – they support economy/create jobs and contribute to the conservation of the environment. The concept of green jobs integrates the social, economic and environmental components of sustainable development. The respect for workers' rights, the increase in productivity and the minimising of negative impacts on the environment in the long run create viable companies and thereby the sustainability of the overall economy (ILO, 2013: 2).

The reform of the global development model should enable dealing with specific issues at national or local level. The importance of stronger and more proactive actions of local development actors is in setting goals for solving problems specific to cities (poverty, waste management, resource utilisation, etc.), which is reflected in proposed SDG 11: make cities and human settlements inclusive, safe, resilient and sustainable, which explicitly includes goals related to sustainable urban development.

Today, about 50 per cent of the world population lives in cities, and it is expected that this share will increase by 2050 to 70 per cent. Urbanisation will be the dominant trend over the next few decades.

Cities are centres of economic activities and environmental pressures, but also of greatest poverty. About 75 per cent of all economic activities are carried out in the cities. Further urbanisation is expected to increase the share of cities in the creation of the world's GDP and investment activities. Due to the growing economic potential of city centres and growing pressures on the quality of life in cities (social inequality, infrastructure and environment), a separate goal of achieving the sustainable development of cities is proposed. This goal should be aimed at achieving the 'green', well planned, productive, safe, inclusive and healthy development of cities. In this sense the UN predicts that the future development of cities should be oriented towards the principles of good governance, environmental policy planning and management, the promotion of economic development based on the exploitation of local resources, strengthening education, cooperation, the research and exchange of knowledge between the cities in developed and underdeveloped countries and the implementation of participatory management model (UN Habitat, 2013).

National approaches

European Union and its member states

Going green and going international

Sustainable development became a fundamental objective of the EU in 1997, when it was included in the Treaty of Amsterdam (and reiterated in the Treaty of Lisbon). Before that, elements of sustainable development were integrated in European environmental protection programmes (see Table 4.4).

The first EU sustainable development strategy was launched by the European Council at the Gothenburg Summit in June 2001. The 2001 strategy was composed of two main parts: the first with objectives and policy measures, and the second calling for a new approach to policy-making. The overall aim of the EU sustainable development strategy was

> to identify and develop actions to enable the EU to achieve a continuous long-term improvement of quality of life through the creation of sustainable communities able to manage and use resources efficiently, able to tap the ecological and social innovation potential of the economy and in the end able to ensure prosperity, environmental protection and social cohesion.
>
> (European Council, 2001)

Table 4.4 EU initiatives towards sustainable development

Year	*Ideas and initiatives*
1992	Fifth EU environmental protection programme 'Towards Sustainability'
1994	Aalborg Charter for Sustainable Cities
1997	Treaty of Amsterdam – sustainable development – one of key objectives
2001	First EU sustainable development strategy (Gothenburg)
2005	Revision of EU sustainable development strategy – integration of sustainability guidelines in all development policies and coordination between EU countries
2009	New revision of EU-SDS to EU enlargement policy towards EU as a key global player in climate change policies
2010	Europa 2020 – European strategy for smart, sustainable and inclusive growth – a priority 'green economy'
2013	EC – A decent life for all: Ending poverty and giving the world a sustainable future – key challenges and opportunities as Rio +20 conference
2015–2020	Seventh environmental action programme of the EU – going green, going international

Source: authors' compilation.

The central instrument of the new approach to policy making was impact assessment, which was expected to ensure that the EU's economic, social and environmental policies mutually reinforced each other. The persistence of unsustainable trends required a revision of sustainable development strategy in 2006. The renewed sustainable development strategy set overall objectives and concrete actions for seven priority areas:

1 climate change and clean energy;
2 sustainable transport;
3 sustainable consumption and production;
4 conservation and management of natural resources;
5 public health;
6 social inclusion, demography and migration;
7 global poverty and sustainable development challenges.

By 2008 the European Strategy on Sustainable Development of Cities had been adopted, based on the Aalborg Charter of 1994.

A revision of the EU sustainable development strategy was again conducted in 2009 due to the observed trend in unsustainable movements. The enlargement underlined the need to strengthen the coordination and clear distribution of responsibilities in the implementation of sustainable development policies.

In accordance with Agenda 21 and the MDGs, the EU sustainable development strategy also requires EU member states to develop NSDSs and stipulates the importance of the external dimension of sustainable development, e.g. the need to tackle global poverty and strengthen cooperation with partners outside the EU. However, sustainable development is still regarded as one of the pillars of growth (first the Lisbon Strategy, and since 2011 the EU 2020 Strategy) instead of an overarching principle.

The EU 2020 Strategy promotes sustainable growth as way to achieve a resource efficient, greener and more competitive economy. In this strategy, sustainable growth means

- building a more competitive low carbon economy that makes efficient, sustainable use of resources;
- protecting the environment, reducing emissions and preventing biodiversity loss;
- developing new markets (domestic and international) for products based on new green technologies and production methods;
- introducing efficient smart electricity grids;
- harnessing EU-scale networks to provide businesses (especially small manufacturing firms) an additional competitive advantage;
- improving the business environment, in particular for SMEs;
- helping consumers make well-informed choices.

To monitor the progress, the 20–20–20 targets for 2020 were defined:

- reducing GHG emissions by 20 per cent compared to 1990 levels;
- increasing the share of renewables in final energy consumption to 20 per cent;
- moving towards a 20 per cent increase in energy efficiency.

Such growth promotes greater energy security (by decreasing import and fossil fuel dependency), competitiveness, protection of the environment (natural resources and climate change) and is supported by two flagship initiatives: a resource efficient Europe; and an industrial policy for the globalisation era.

Based on the principle of subsidiarity, all EU member states have prepared NSDSs.[2] Bulgaria has not officially adopted the draft prepared, while Poland announced in 2007 that its strategy (adopted in 2000) was outdated. In line with the EU's economic governance structure, member states also prepare national reform programmes (NRPs), which define the national targets of the Europe 2020 Strategy.

The *United Kingdom* first developed an NSDS in 1994, and revised it in 1999 and again in 2005. In February 2011, the UK government published a policy paper entitled 'Mainstreaming sustainable development – the Government's Vision and what this Means in Practice', which presents measures to deliver sustainable development through the green economy, action to tackle climate change, protecting and enhancing the natural environment, fairness and improving well-being, and building a big society (DEFRA, 2011). The progress on implementation is monitored every year. The OECD identified the UK's integrated policy appraisal as an example of best practice. In implementing its sustainable development strategy, the United Kingdom integrated its policy appraisal approaches into the Cabinet Office's Regulatory Impact Assessment process to require departments to assess the sustainability impacts of policy proposals and link this to mainstream public consultation processes.

Since 2013, sustainable development is considered part of policy development, and also considered in all activities of the government (i.e. how it runs its buildings and how it buys its goods and services). All departments are responsible for making sure that their own policies and activities contribute to sustainable development. DEFRA has a role in overseeing sustainable development across central government.

The main sustainability challenges according to the UK's 2014 NRP relates to the sustainability of public finances and sustainable employment (especially for youth).

In *Denmark* the previous Danish government published the latest NSDS in 2009. It centres around three core principles:

- to commit and motivate all actors to take responsibility for a sustainable development;
- to develop innovative and environmentally friendly solutions; and
- to take long-term global consequences into account.

The Strategy for Sustainable Development brings forward goals and specific initiatives for the following nine areas:

- globalisation as a benefit for all;
- climate change;
- nature for the future;
- green innovation in production and consumption;
- well-functioning urban areas;
- better health for all;
- knowledge, research and education in an innovative society;
- man as a resource;
- responsible, long-term and sustainable economic policy.

Despite announcements of the new government in its programme (October 2011) that a new and binding sustainability strategy with fixed goals, time limits and associated indicators and monitoring processes will be developed, the 2009 strategy is still the latest national version. Monitoring and reporting are coordinated by the Danish Environmental Protection Agency, although there are no exact deadlines or exact formats for reports on the strategy.

At the Nordic level, the strategy 'A Good Life in a Sustainable Nordic Region' was published in 2013. It provides long-term guidelines until 2025 and a cross-sectoral framework for the work of the Nordic Council of Ministers. The strategy focuses on the following areas: the Nordic welfare model, viable ecosystems, the changing climate, sustainable use of the earth's resources, and education, research and innovation. The sectoral ministerial councils within the Nordic Council of Ministers will supplement the strategy with concrete measures, and development in the Nordic region will be monitored using indicators (Nordic Council of Ministers, 2013).

The United States of America

Achieving sustainability through the modernisation of global cooperation

President Bill Clinton established a President's Council on Sustainable Development in 1993 to recommend a National Action Strategy for sustainable development. By early 1996 the council reached agreement (President's Council on Sustainable Development, 1996) on a set of ten national goals:

1. Health and the environment (ensure that every person enjoys the benefits of clean air, clean water, and a healthy environment at home, at work, and at play).
2. Economic prosperity (sustain a healthy US economy that grows sufficiently to create meaningful jobs, reduce poverty, and provide the opportunity for a high quality of life for all in an increasingly competitive world).

3 Equity (ensure that all Americans are afforded justice and have the opportunity to achieve economic, environmental, and social well-being).
4 Conservation of nature (use, conserve, protect and restore natural resources of land, air, water and biodiversity in ways that help ensure long-term social, economic and environmental benefits for ourselves and future generations).
5 Stewardship (create a widely held ethic of stewardship that strongly encourages individuals, institutions and corporations to take full responsibility for the economic, environmental and social consequences of their actions).
6 Sustainable communities (encourage people to work together to create healthy communities where natural and historic resources are preserved, jobs are available, sprawl is contained, neighbourhoods are secure, education is lifelong, transportation and health care are accessible and all citizens have opportunities to improve the quality of their lives).
7 Civic engagement (create full opportunity for citizens, businesses and communities to participate in and influence the natural resource, environmental and economic decisions that affect them).
8 Population (move toward stabilisation of the US population).
9 International responsibility (take a leadership role in the development and implementation of global sustainable development policies, standards of conduct, and trade and foreign policies that further the achievement of sustainability).
10 Education (ensure that all Americans have equal access to education and lifelong learning opportunities that will prepare them for meaningful work, a high quality of life, and an understanding of the concepts involved in sustainable development).

In 1996 the director of sustainable development and the Council for Sustainable Development were created in the US Department of Agriculture (USDA). However, despite the fact that the USDA is formally charged for sustainable development, the Bureau of Oceans and International Environmental and Scientific Affairs (2011) prepared a submission for the Rio+20 Conference. It focused on three key issues, relevant for achieving sustainability:

- the built environment (clean energy and urbanisation);
- the natural environment (ecosystems management and rural development); and
- the institutional environment (modernising global cooperation).

Despite growing awareness of the need to build a sustainable national economy, and the promotion of sustainable practices in external action, the US economy is characterised by linear flows of materials, heavy dependence on fossil fuels, limited use of renewable resources, and resource use that is strongly connected to economic growth.

Canada

Greening the government's operation

Until 2010 the principle of sustainable development was included in sectoral strategies, and there were no actions at the federal level. Each department prepared its own sustainable development strategies for the three-year period.

The first federal sustainable development strategy (FSDS) for the period 2010–2013, was tabled in Parliament in October 2010, based on the 2008 Federal Sustainable Development Act.

The Federal Sustainable Development Act provides the legal framework for developing and implementing an FSDS that will make environmental decision making more transparent and accountable to Parliament. It defines the basic principle that sustainable development is based on an ecologically efficient use of natural, social and economic resources, and acknowledges the need to integrate environmental, economic and social factors in the making of all decisions by government.

The Act required that within two years after its entry into force and within every three-year period after that, the minister shall develop an FSDS based on the precautionary principle.

The FSDS sets out federal SDGs and targets and an implementation strategy for meeting each target and identifies the minister responsible for meeting each target.

The first cycle of the FSDS was adopted in 2010. Since 2011, departments and agencies have produced annual departmental sustainable development strategies, integrated into their core planning and reporting processes.

The second cycle of the FSDS builds on three key improvements made in the first cycle:

1 an integrated, whole-of-government picture of actions and results to achieve environmental sustainability;
2 a link between sustainable development planning and reporting and the government's core planning and reporting processes; and
3 effective measurement, monitoring and reporting in order to track and report on progress to Canadians.

The first cycle of FSDS increased transparency and provided a comprehensive picture of actions across the federal government that contribute to environmental sustainability. Since 2011, departments and agencies have produced annual departmental sustainable development strategies, integrated into their core planning and reporting processes.

The key steps for the second cycle (2013–2016) include:

1 building the whole-of-government picture by incorporating a broader range of federal government actions that support the FSDS goals and targets;

2 enhancing the link to core federal planning and reporting through ongoing alignment between the FSDS and federal departments' Program Alignment Architectures; and
3 expanding the Canadian Environmental Sustainability Indicators suite to ensure that indicators are available to measure progress on all FSDS goals and targets.

Further development of the government's sustainable development approach will continue in future FSDS cycles. Like the first cycle, the second cycle of the FSDS includes four priority themes:

1 addressing climate change and air quality;
2 maintaining water quality and availability;
3 protecting nature and Canadians; and
4 shrinking the environmental footprint – beginning with government.

The first three themes reflect the government's priorities for environmentally sustainable policies and programmes. The fourth theme focuses on shrinking the federal government's environmental footprint.

The themes included are aspirational long-term goals and medium term targets that are specific, measurable, achievable, relevant, and time-bound, and are concrete implementation strategies.

The 2013–2016 FSDS includes eight goals across the four themes, supported by 34 targets. In total, the strategy includes 225 implementation strategies, which are programmes and activities undertaken by departments across the federal governments.

Australia

Sustainable development improves total quality of life

The National Strategy for Ecologically Sustainable Development provides broad strategic directions and framework for governments to direct policy and decision making. It was adopted by the Council of Australian Governments in 1992. This strategy has evolved over several years based on the World Conservation Strategy in 1980, the National Conservation Strategy for Australia in 1983, and the 1987 report of the World Commission on Environment and Development – Our Common Future (the Brundtland Report).

The goal of the strategy is 'development that improves the total quality of life, both now and in the future, in a way that maintains the ecological processes on which life depends'. The core objectives are:

- to enhance individual and community well-being and welfare by following a path of economic development that safeguards the welfare of future generations;

- to provide for equity within and between generations;
- to protect biological diversity and maintain essential ecological processes and life-support systems.

The strategy defines ecologically sustainable development as: 'using, conserving and enhancing the community's resources so that ecological processes, on which life depends, are maintained, and the total quality of life, now and in the future, can be increased' (Ecologically Sustainable Development Steering Committee, 1992).

Mandated by the Environment Protection and Biodiversity Conservation Act 1999, Australian government departments and agencies (Department for Environment, National Water Commission, Director of National Parks, Great Barrier Reef Marine Park Authority, Supervising Scientist, Bureau of Meteorology, Murray-Darling Basin Authority, Sydney Harbour Federation Trust, Climate Change Authority and Clean Energy Regulator) are required to report on their sustainability performance as part of their annual reports.

The Department of the Environment, Water, Heritage and the Arts is the lead government agency for developing and implementing national policy, programmes and legislation to protect and conserve the natural environment. One of the key functions of the department is to promote and support ecologically sustainable development. Its latest report, for the 2013–2014 period, provides an overview of activities and achieved results related to:

- biodiversity and ecosystems;
- sustainable population and communities;
- Antarctica;
- water;
- environmental protection and heritage conservation;
- commonwealth environmental water; and
- climate change.

The main activities regarding biodiversity and the ecosystem relate to the support of the Great Barrier Reef, one of the world's greatest natural wonders, and implementing the system of threat abatement planning. The work started on the development of the Reef 2050 Long-term Sustainability Plan, an overarching framework that will guide the protection and management of the World Heritage Area from 2015 to 2050.

In the area of sustainable population and communities, activities relate to the preparation of a National Clean Air Agreement, to be in place by July 2016, the review of the Ozone Protection and Synthetic Greenhouse Gas Management Act 1989 and related legislation, as well as the review of the Product Stewardship (Oil) Act 2000, and relevant provisions of customs and excise legislation, and the implementation of the National Television and Computer Recycling Scheme.

Regarding Antarctica, the preparation of a 20 year strategic plan for Australia's Antarctic operations was launched, as well as the process of procuring a new icebreaker.

In the water sector, water reform in the Murray–Darling Basin was launched. All basin jurisdictions are working collaboratively to develop an agreed package of supply, efficiency and constraints measures for consideration under the Sustainable Diversion Limit adjustment mechanism by mid-2016. Also, the Water Recovery Strategy for the Murray–Darling Basin was launched in 2014, outlining the future approach for water recovery for the environment. The government is giving priority to investment in water savings infrastructure projects over purchasing as a means of returning water to the environment.

In 2013, the Australian government announced that the Australian Heritage Strategy would form one of Australia's key heritage priorities. The draft Australian Heritage Strategy was released for consultation in 2014, and the strategy is expected to be finalised in 2015.

Improving the health of rivers is a long-term process. Monitoring results of the use of Commonwealth environment water indicates positive impacts on the basin environment, such as the improved health of native fish populations (including the breeding of Golden Perch in Victoria's Goulburn River for the first time in ten years, and supporting Murray Cod breeding in Gunbower Creek), improved lake levels and water quality, including improved salinity in the Lower Lakes and Coorong in South Australia, and increased environmental resilience.

In 2013–2014, the Murray–Darling Basin Authority published the first set of basin annual environmental watering priorities. Commonwealth environmental water contributed to eight of the ten priority outcomes.

Activities regarding climate change include the carbon tax repeal legislation and the Emissions Reduction Fund White Paper.

Thus, despite the fact that the sustainable development strategy has not been updated, the pursuit of ecologically sustainable development has been increasingly incorporated into the policies and programmes of Australian governments as a significant policy objective.

Work at the national level is undertaken within the context of the:

- Intergovernmental Agreement on the Environment, and the
- Heads of Agreement on Commonwealth/State Roles and Responsibilities for the Environment.

These agreements establish arrangements by which Commonwealth, state/territory and local governments interact on environmental matters.

Japan

A new growth through green innovation

Japan does not have a separate NSDS, having defined its own model of a sustainable society, which is based on a low carbon economy, sound material cycle and biodiversity conservation (OECD, 2010).

Various studies consider specific strategic plans as sustainable development plans. For instance, the Sustainable Development Knowledge Platform considers the New Growth Strategy as a sustainable development strategy, the World Bank considers Japan's energy efficiency strategy as an example of a national green development strategy (World Bank, 2012: 234). National Agenda 21 Action Plans, aimed at the implementation of Agenda 21 (including its sustainability dimension), can be also considered as sustainable development implementation plans. These include the National Agenda 21 Action Plan (submitted to the UN in 1994) and the Basic Environmental Plan.

The first Basic Environment Plan was developed in 1994, the second in 2000, the third in 2006 and the fourth in 2012. These plans outline the government's comprehensive and long-term measures on environmental policy and related policy issues. The latest plan tries to reconcile Japan's environmental goals (conserve biodiversity, clean up water and air) with the new reality that the country faces in the wake of the Fukushima nuclear disaster (decentralise and diversify energy sources, take measures to clean up disaster debris and implement plans for cleaning nuclear waste). The document is sprinkled with ambitious targets, such as raising JPY50 trillion (US$626 billion) in new environment related businesses and 1.4 million environment-related jobs by 2020. But the target of cutting global-warming gases to 80 per cent of 1990 levels by 2050 probably raises the most scepticism.

Energy efficiency strategy, whose implementation is guided by the 2006 'Energy Conservation Law', set targets to improve energy efficiency by 30 per cent by 2030 relative to 2006. Japan's energy intensity decreased by 26 per cent between 1980 and 2009; it is one of the most energy-efficient countries in the world. Fostering energy conservation technologies and development a benchmarking approach to monitoring energy conservation are the main instruments for further improvements of energy efficiency. Japan's Top Runner Programme provides for setting the benchmarks. It tests 21 types of appliances, ranging from vending machines and air conditioners to television sets. Test results are used to determine the most efficient model, whose level of efficiency is set as the new baseline. Then, manufacturers have the obligation to achieve the new baseline within four to eight years. In 2010, Japan introduced the innovative concept of a 'smart community', which maximises the use of renewable energy in cities and relies on smart grids to deal with its intermittent nature. In addition, the introduction of integrated energy consumption standards for all buildings and the target of net-zero-energy houses by 2020 (and the norm nationwide by 2030) are planned.

The New Growth Strategy provides for 'green innovation' in the fields of environment and energy. This strategy aims to:

- create over JPY50 trillion in new environment-related markets;
- create 1.4 million new environment sector jobs;
- reduce worldwide GHG emissions by at least 1.3 $GtCO_2e$ (equivalent to the total emissions of Japan), using Japanese private sector technology.

Republic of Korea

The ROK has continually implemented strategies for sustainable development since the Rio Summit in 1992, when the first NSDS was prepared to implement Agenda 21. Several presidents highlighted their commitments to sustainable development:

- Kim Young-Sam (the fourteenth president, 1993–1998), pronounced that he would be the president to improve the environment.
- Kim Dae-Jung (the fifteenth president, 1998–2003) declared an environmental vision for the new millennium. In 2000 the Presidential Commission on Sustainable Development (PCSD) was established. It has developed various sectoral-based strategies for sustainable development (energy, water, coastal waters, land, climate change, transportation, conflict management and official development aid).
- Ro Mu-Hyun (the sixteenth president, 2003–2008) in 2005 declared 'A National Vision for Sustainable Development'. The vision defined as a goal to continue Korea's growth as 'an advanced country while maintaining balance between the economy, society and the environment'.
- Lee Myung-bak (the seventeenth president, 2008–2013) established the Presidential Commission on Green Growth that replaced the PCSD.

There is a well-established framework for long-term planning, both cross-sectoral and sectoral, and with a high degree of spatial diversification (planning at national, regional and local levels). The national vision for sustainable development provided a framework for governmental departments to participate in producing the NSDS and in formulating implementation plans/action programmes that can actualise the National Sustainable Development Vision.

Based on the vision, an NSDS and appropriate action programmes were adopted in 2006 (ROK, 2006). There are five main themes:

- sustainable management of natural resource;
- social integration and national health promotion;
- sustainable economic development;
- dealing with climate change and global environmental issues; and
- institutionalisation and education for sustainable development.

Each theme has numerous key action programmes. Each action programme contains a detailed list of objectives and actions, centred on specific issues. The NSDS was developed for the period until 2010, when it was replaced by Green Growth strategy and its implementation action plan. The first Five-Year Action Plan of the Green Growth strategy includes a four-step strategy:

1 promote strategic R&D and commercialisation of NRE technologies;
2 create new market to stimulate industrialisation of NRE field;

3 facilitate export industrialisation; and
4 strengthen growth potential by providing infrastructure building for private sectors working in NRE technologies.

Together with promoting green growth as a means to implement a sustainable development strategy at a national level, the ROK supported the establishment of a United Nations Office for Sustainable Development in 2011. The office is managed by the UN Department of Economic and Social Affairs (UNDESA) and serves as a research and training facility. It is designed to help poorer countries pursue sustainable development to accelerate economic growth while improving the quality of life and protecting the environment, launched by the UN, the ROK's environment ministry, the city of Incheon and Yonsei University.

China

Sustainable growth through upgraded industries and domestic consumption

Following the outcome of the United Nations Conference on Environment and Development in Rio de Janeiro in 1992, China started drafting its national Agenda 21. In 1994, China released its Agenda 21 – the White Paper on China's Population, Environment, and Development in the 21st Century – as a strategic document for guiding the country's social and environmental development. In 1996, China incorporated sustainable development into national strategies and began its full implementation.

China's Agenda 21 marked the beginning of the sustainable development process in China. It deals with:

1 general strategic issues of sustainable development;
2 societal aspects of sustainable development;
3 sustainable development of the economy; and
4 protection of resources and the environment.

Since the turn to a new century, China has further deepened its understanding of the connotations of sustainable development. To facilitate implementation and review the progress achieved, the Programme of Action for Sustainable Development in China in the Early Twenty-first Century was adopted in 2002. As its main achievements, the programme identifies economic growth (providing an improvement of people's livelihood and quality of life); the reduction of population growth and the narrowing of the East–West gap, the increase of government expenditures for ecological protection and a growing public awareness of SDGs through capacity building. The greatest challenges are the conflict between rapid economic growth, intensive resource extraction and related ecological damages, regional socio-economic disparities, the aging of the population, pressure through unemployment and inconsistencies between existing law and requirements for sustainable development.

In 2003, China proposed the scientific outlook on development with people-oriented, comprehensive, coordinated and sustainable features. The scientific outlook on development is a major theory put forward by China that, based on its own realities, summarises development practices, draws on foreign experience and adapts to new development situations at home and abroad. Since 2003, China has proposed advanced concepts such as a resource-saving and environment-friendly society, an innovation-oriented country, ecological civilisation and green development, and has continuously put them into practice.

Progress achieved in the economic restructuring and transformation of development patterns can be illustrated by the developing agriculture, which led to increased food security; a new path to industrialisation; shifting the mode of economic development by adjusting industrial structure; developing a circular economy; promoting the development of strategic and emerging industries; and upgrading traditional industries. China also uses information technology to promote industrialisation and to enhance the competitiveness of the manufacturing sector. China became the first country that achieved the MDGs of halving the number of people living in poverty. Efforts have been made in China to realise free compulsory education, improve healthcare capabilities and service levels, accelerate the construction of a social security system covering both urban and rural residents, further protect the rights and interests of women and children, enhance people's overall quality of life and bring into shape the quality human resources that can support sustainable development. Substantive results have been achieved in resource conservation and environmental protection. Air quality has improved, as well as energy security. China implements strict arable land and water resources protection systems and has developed a system for marine protected areas. The large-scale ecological restoration drive has been underway, with the principle of 'overall progress and breakthroughs in key points', which increased forest coverage from 16.55 per cent in 2000 to 20.36 per cent in 2010. China is trying to explore a rapid path for industrialisation and urbanisation that features the efficient use of natural resources, the protection of the ecological environment, and coordination between socioeconomic development and the environment and natural resources.

The concept of sustainable development is also integrated into the five-year planning process of China's economy. The five-year plan is a strategic national roadmap, setting priorities regarding China's future socioeconomic development, and providing guidelines and targets for policy making at the sectoral and sub-national level. The current five-year plan (2011–2015) emphasises three main priorities: sustainable growth, industrial upgrading and the promotion of domestic consumption. These priorities explain why certain sectors, including energy, automotive, IT infrastructure and biotechnology, also receive a high degree of focus.

Mexico

Sustainable development concept is in Mexico's constitution

Mexico does not have an overarching, cross-sectoral NSDS, but the concept of sustainable development is included in the constitution and major national policy strategies and goals.

Mexico's main planning instrument is its NDP, currently covering the period 2013–2018. The NDP outlines the main objectives and strategies of public policy and defines indicators. It is the central document for the whole federal public administration and is legally approved by the Mexican Congress. Sectoral programmes from the different ministries that also include sustainability elements support the NDP. The current NDP defines five main goals:

1 a peaceful Mexico;
2 an inclusive Mexico;
3 a Mexico with quality education;
4 a prosperous Mexico; and
5 a Mexico with global responsibility.

Under the current NDP, the commitment of sustainable development is part of the overarching goal related to prosperity. It includes an economic, environmental and social dimension: to preserve macroeconomic stability, expand access to financing, promote greater employment and productivity with equality; support sustainable green growth to preserve natural resources and generate wealth, improve competitiveness and employment; promote better and broader access to telecommunications services, and provide sufficient energy at competitive prices.

Mexico is also actively involved with other South and Latin American countries on a variety of sustainable development initiatives, such as the Latin American and Caribbean Initiative (ILAC) for sustainable development.

ILAC was approved in 2002 by the Forum of Ministers of Latin America and the Caribbean at the World Summit on Environment and Sustainable Development in Johannesburg.[3] ILAC is part of the implementation plan adopted at the aforementioned summit. It provides the common agenda of Latin American and Caribbean countries, which has been fostered through the regional action plans (RAPs) adopted by the Forum of Ministers. The RAPs from 2004–2005 and 2006–2007 specified regional priorities and lines of action. Based on ILAC's main results, the revision was launched in 2007 and new priority action areas were identified in 2008, related to:

1 biological diversity;
2 management of water resources;
3 vulnerability, human settlements and sustainable cities;
4 social issues, including health, inequity and poverty;

5 economic aspects, including competitiveness, trade and the production and consumption patterns (energy); and
6 institutional aspects.

Brazil

Sustainability through savings

The Brazilian Agenda 21 was adopted in July 2002 in preparation for the World Summit on Sustainable Development in Johannesburg. It is considered as a social pact, as it was created through years of extensive consultation (1997–2002) and was not adopted as an official document. The Brazilian Agenda 21 defines 21 objectives organised in the following four priority areas:

1 the economy of savings in the knowledge society;
2 social inclusion for a solidarity society;
3 strategy for urban and rural sustainability;
4 strategic natural resources: water biodiversity and forests.

The relation of the Agenda 21 to the government policies is mainly through its link to the multi-year NDP. Under the Brazilian constitution, the government's multi-year national development strategy must be prepared every four years and approved by the National Congress. It defines the priorities for the social areas, investment on infrastructure and in income generating sectors. The current multi-annual plan covers the period 2012–2015.

Brazil is also actively involved with other South and Latin American countries on a variety of sustainable development initiatives, such as the ILAÇ.

Turkey

High economic growth, strong society and sustainable development

Development policies in Turkey have progressed towards sustainable development in parallel with international development. The development plan is the fundamental policy document outlining the economic, social and cultural reforms mandatory for the public sector and the directive for private sector. Since the 1990s development plans have included sustainable development issues.

The sixth development plan (1990–1994), among elements of economic and social development, identified the protection of natural resources and the environment. The seventh development plan (1996–2000) included policies that give priority to the integration of the environment and the economy in accordance with the principles of sustainable development. In the eighth development plan (2001–2005), a sustainable development approach was adopted, based on protecting human health, ecological balance and historical and aesthetic values while realising economic and social development. The ninth development plan

(2007–2013) that followed, integrated sustainable development principles into the development targets. The current development plan, the tenth development action plan (2014–2018), is based on the 2023 goals the government announced as an election manifesto in 2011, right before the general elections. In 2023, the Republic of Turkey will celebrate its centenary. Its economic goal is to rank among the world's ten biggest economies. To move from its current seventeenth position, Turkey will have to grow on by average 8 per cent per annum.

The plan expresses its vision as follows:

> The Tenth Development Plan that covers the 2014–2018 period will be a milestone in the country's endeavour to enhance the current level of welfare in line with the 2023 goals. The Tenth Development Plan is designed to cover issues such as the rule of law, information society, international competitiveness, human development, environmental protection and sustainable use of resources as well as high, steady and inclusive economic growth.

It presents the objectives in the light of four dimensions of development:

1 qualified people, strong society;
2 innovative production, steady high growth;
3 liveable places, sustainable environment; and
4 global partnership for development (Cağlar, Acar, 2013).

The plan foresees creating 4 million new jobs with a 2.9 per cent annual increase. The labour force participation rate is aimed to increase to 53.8 per cent by the end of 2018, from 50 per cent in 2012. A 2 per cent decrease in the unemployment rate is targeted, as the plan envisages decreasing the unemployment rate from 9.2 per cent in 2012 to 7.2 per cent (*Hürriyet Daily News*, 2013).

The sustainable development approach that was integrated into the development plan takes place in many sectoral and thematic strategy documents (see Box 4.1).

Box 4.1 Selected Turkish national action plans including sustainable development

National Environment Strategy and Action Plan

1998–2018 – To increase quality of life, develop environmental awareness and sensitivity, improve environmental management and provide sustainable economic, social and cultural development.

Turkey's National Action Programme to Combat Desertification

2005–2015 – To prevent and/or reduce the effects of desertification and drought by determining the factors that cause desertification.

Tourism Strategy of Turkey

2007–2023 – To increase employment and make tourism a leading sector in regional development in a sustainable manner.

EU Integrated Environmental Approximation Strategy

2007–2023 – To create a healthy and balanced environment in which the basic needs of present and future generations are met, quality of life is increased, biological diversity is protected and natural sources are managed in a rational perception with a sustainable development approach.

National Biological Diversity Strategy and Action Plan

2008–2017 – To protect and sustain the use of biological diversity and genetic resources.

Waste Water Treatment Action Plan

2008–2012 – To build waste water treatment facilities in accordance with the country's needs and priorities, to maximise the benefits of waste water investments, and to optimise the basin management by reusing treated waste water.

Waste Management Action Plan

2008–2012 – To manage domestic and industrial waste in accordance with the principle of sustainability.

National Afforestation Action Plan

2008–2012 – To sequestrate 181.4 $MtCO_2eq$ from the atmosphere into forest in 20 years by afforestation and breeding.

Lifelong Learning Strategy and Action Plan

2009–2013 – To increase people's employment capabilities in line with the changing and developing economy and labour force market.

Integrated Urban Development Strategy and Action Plan

2010–2023 – To strengthen the economic, social and cultural structures of the settlements via an integrated approach to the subjects of planning, settlement and structuring on a national level, and to increase the liveability levels and qualities of places and lives.

National Climate Change Strategy

2010–2020 – To contribute to the combat with climate change in accordance with the sustainable development policies within the frame of common but differentiated responsibilities, principles and national circumstances.

National Rural Development Strategy and Rural Development Plan

2010–2013 – To improve working and living conditions of rural people locally in accordance with urban areas and to ensure sustainability, by utilising local potential and resources and protecting natural and cultural values.

The Strategy and Action Plan for Strengthening Employment and Vocational Training

2010 – To provide vocational and technical training in line with the labour market needs, to strengthen the relationship between education and employment, to efficiently implement active labour market policies, to eradicate the problem of not having any profession and to increase the skills and abilities of the labour force to be employed.

Climate Change Action Plan

2011–2023 – To limit GHG emission in the combat with climate change, to adapt to climate change and minimise the effects of climate changes.

National Science, Technology and Innovation Strategy

2011–2016 – To increase the progress accomplished in improving R&D and innovation capacity in recent years and to make it sustainable.

Turkish Industrial Strategy Document

2011–2014 – To increase the competitiveness of the industry and accelerate the transformation into an industrial structure that takes a greater share from world exports, in which high value added and high technology goods are produced and that has more qualified labour and at the same time is more sensitive about the environment and society.

SME Strategy and Action Plan

2011–2013 – To enhance the production level, amount of investment, value added and growth of SMEs.

Care Services Strategy and Action Plan

2011–2013 – To increase the quality of care services provided to the disabled and to ensure the sustainability of these services.

Energy Efficiency Strategy Document

2012–2023 – To reduce the energy intensity of the economy with energy efficiency and thus strengthen supply security in energy, to reduce the risks of import dependency, to minimise costs, to combat climate change and to protect the environment.

National Earthquake Strategy and Action Plan

2012–2023 – To prevent or reduce physical, economic, social, environmental and political damage and losses caused by earthquakes.

At a regional level, as a party to the Barcelona Convention, Turkey participates in the Mediterranean Strategy for Sustainable Development (MSSD). The MSSD was adopted in 2005 and provides guidance for national decision makers to address sustainable development issues, implement international agreements and initiate partnerships.

The strategy pursues four main directions:

1 to contribute to economic development while building on Mediterranean assets;
2 to reduce social disparities and fulfil MDGs while strengthening diversity;
3 to ensure the sustainable management of natural resources and change consumption and production patterns;
4 to improve governance at local, national, regional levels.

In order to reach these four objectives, the MSSD identifies seven priority fields of action: water, energy, transport, tourism, agriculture, urban development, and sea and coastal management.

Based on an assessment of the implementation of the MSSD for the period 2005–2010, a revision of the MSSD in 2015 is planned (so called 'MSSD 2.0'). The steering committee has proposed six thematic areas:

1 seas and coasts: including efforts towards reaching 'Good' environmental status of Mediterranean marine and coastal ecosystems;
2 natural resources, rural development and food: including links not only to marine and coastal, but also terrestrial habitats, and to fresh waters;
3 climate: including the impacts of climate change on natural resources and socioeconomic sectors and possible responses;
4 sustainable cities: including coastal urbanisation and Integrated Coastal Zone Management, land transport, waste management, infrastructure, housing and energy;
5 transition towards a green economy: including sustainable consumption and production;
6 governance, including financing: Aarhus Convention–related issues such as public participation, and cooperation (Uras, 2014).

However, an assessment of the implementation indicates that the MSSD is a useful document that acts as a background for public authorities in the implementation of national strategies but that it does not affect the policy action in single countries in concrete terms, as the document is lacking detailed suggestions on the modalities by which single countries should incorporate MSSD by national strategies. Thus, the MSSD has had little observable influence on sectoral strategies, policies and action programmes (UNEP/MAP, 2011).

Final remarks

Sustainable development integrates economic, social and environmental objectives. It recognises that economic growth is necessary, but not sufficient for development.

The strategic policy frameworks that include sustainable development issues (e.g. generic NSDSs, sustainable development included in other policy strategies – general development or environmental strategies, etc.) vary significantly among the countries. The strategic policy frameworks largely depend on the respective political-administrative systems, economic development, environmental conditions, institutional capacities, political culture, etc.

Generally, the UN is an important driver for addressing sustainable development issues. It has developed guidelines for NSDSs, and supports the creation and implementation of such strategies, especially in developing countries.

Generic NSDSs can only be found in a limited number of countries. Several other countries have adopted national Agenda 21 strategies (e.g. China, Brazil). Nevertheless, sustainable development is increasingly included in strategic policy documents, most often in NDPs (e.g. Mexico). Environmental policy plans are still dominant in several countries, some of them also address cross-sectoral issues (e.g. Japan).

Despite efforts to address the economic, social and environmental issues in sustainable development strategies, environmental sustainability is in many countries the prevailing approach.

Some regional commitments to develop NSDSs or integrated sustainable development approaches have emerged in several NSDSs world regions, such as the Mediterranean (the MSSD) or the ILAC.

Although sustainable development is increasingly taken up at the national level, major challenges remain, e.g. concrete actions on the ground, integrating sustainable development into all policy sectors, coordination of sustainable development objectives among government levels as well as developing comprehensive and effective monitoring and evaluation processes.

Notes

1 These include: Agenda 21, the strategy for sustainable policies in the twenty-first century, the Rio Declaration on Environment and Development, which states that the only way to long-term economic progress is to link it with environmental protection and health, Forest Principles, which deal with sustainable use of forests and balance the need to protect forests (for environmental and cultural reasons) and to use trees and other forest life for economic development, the Convention on Biological Diversity, aiming at sustainable use of biological diversity, and the Framework Convention on Climate Change, which sets long-term objectives to stabilise atmospheric GHG concentrations at a level that would prevent dangerous anthropogenic interference with the climate system and an overall framework to tackle challenges posted by climate change.

2 For more information see the Sustainable Development Network website, online, available at: www.sd-network.

3 Participating countries include Brazil, Cuba, Chile, Ecuador, El Salvador, Guatemala, Mexico, Panama, the Dominican Republic and the Bolivarian Republic of Venezuela.

References

Amnesty International, 2012. 'Nigeria: Oil Spill Investigations a Fiasco in the Niger Delta'. Online, available at: www.amnesty.org/en/news/nigeria-oil-spill-investigations-fiasco-niger-delta-2012-08-02 (accessed 4 December 2014).

Anderson, S. and Cavanagh, J., 2000. The Rise of Corporate Global Power. Washington, DC: Institute for Policy Studies.

Benson, E. and Greenfield O., no date. 'Post Rio to Post-2015 Think Piece, Surveying the "Green Economy" and "Green Growth" Landscape'. UNEP. Online, available at: www.unep.org/civil-society/Portals/24105/documents/NY%20consultation/GE%20thinkpiece.pdf (accessed 27 November 2014).

Cağlar E. and Acar, O., 2013. 'An Assessment on the 10th Development Plan'. Evaluation Note No. 201330, Economic Policy Research Foundation of Turkey. Online, available at: www.tepav.org.tr/upload/files/1378461158-4.An_Assessment_on_the_Tenth_Development_Plan.pdf (accessed 30 October 2014).

Chung, Y.-K. and Hwan, K., 2006. 'The Korean National Strategy for Sustainable Development: A Background Report'. Online, available at: http://sustainabledevelopment.un.org/content/documents/1394backgroundReport.pdf (accessed 21 November 2014).

Constanza, R. and Kubiszewski, I., 2014. 'Creating a Sustainable and Desirable Future'. Sydney: World Economic.

Danish Energy Agency, 2012. 'Green Production in Denmark – and its Significance for the Danish Economy'. Online, available at: www.ens.dk/sites/ens.dk/files/policy/green-production-denmark-contributes-significantly-danish-economy/Green%20production%20in%20Denmark%20-%20web%20111212.pdf (accessed 21 November 2014).

DEFRA, 2011. 'Mainstreaming Sustainable Development – the Government's Vision and what this Means in Practice'. Online, available at: www.gov.uk/government/uploads/system/uploads/attachment_data/file/183409/mainstreaming-sustainable-development.pdf (accessed 4 October 2014).

EC, 2013. 'A Decent Life for All: Ending Poverty and Giving the World a Sustainable Future', COM (2013) 92 final, Brussels, 27.2.2013.

Ecologically Sustainable Development Steering Committee, 1992. 'National Strategy for Ecologically Sustainable Development'. Council of Australian Governments. Online, available at: www.environment.gov.au/about-us/esd/publications/national-esd-strategy (accessed 6 December 2014).

Environmental Leader, 2009. 'Shell Scales Back Renewable Energy Investments'. Online, available at: www.environmentalleader.com/2009/03/18/shell-scales-back-renewable-energy-investments/ (accessed 21 November 2014).

Fortune, 2013. Global 500. Online, available at: http://money.cnn.com/magazines/fortune/global500/2013/full_list/?iid=G500_sp_full (accessed 6 December 2014).

Hürriyet Daily News, 2013. 'New Plan Targets 5.5 Percent Growth'. Ankara: Anatolia News Agency. June. Online, available at: www.hurriyetdailynews.com/new-plan-targets-55-percent-growth.aspx?pageID=238&nid=48987 (accessed 2 November 2014).

ILO, 2013. 'Sustainable Development, Decent Work and Green Job's. Report V. Online, available at: www.ilo.org/wcmsp5/groups/public/--ed_norm/--relconf/documents/meeting-document/wcms_207370.pdf (accessed 4 October 2014).

Local Governments for Sustainability (ICLEI), 2014. 'The World Needs an "Urban" Sustainable Development Goal to Reach Global Sustainability'. Online, available at: www.iclei.org/details/article/the-world-needs-an-urban-sustainable-development-goal-to-reach-global-sustainability.html (accessed 11 November 2014).

Ministry of Development, 2012. 'Turkey's Sustainable Development Report: Claiming the Future'. Ankara. Online, available at: http://sustainabledevelopment.un.org/content/documents/853turkey.pdf (accessed 11 November 2014).

Nordic Council of Ministers, 2013. 'Good Life in a Sustainable Nordic Region: Nordic Strategy for Sustainable Development'. Copenhagen: Nordic Council of Ministers. Online, available at: http://norden.diva-portal.org/smash/get/diva2:701472/FULLTEXT01.pdf (accessed 11 November 2014).

OECD, 2011. *Towards Green Growth.* Online, available at: www.oecd.org/greengrowth/48224539.pdf (accessed 24 September 2014).

Overseas Development Institute (ODI), German Development Institute (DIE), Centre for Development Policy Management (ECDPM), European Report on Development, 2013. 'Post 2015: Global Action for an Inclusive and Sustainable Future', Full Report, Brussels.

Political Economy Research Institute, 2013. 'Toxic 100 Air Polluters Press Release'. Amherst: University of Massachusetts. Online, available at: www.peri.umass.edu/toxic_press/ (accessed 1 November 2014).

President's Council on Sustainable Development, 1996. 'Sustainable America: a New Consensus for Prosperity, Opportunity, and Healthy Environment for the Future'. Online, available at: http://clinton2.nara.gov/PCSD/Publications/TF_Reports/amer-top.html.

President's Council on Sustainable Development, 1999. 'Towards A Sustainable America: Advancing Prosperity, Opportunity, and a Healthy Environment for the 21st Century'. Online, available at: http://clinton2.nara.gov/PCSD/Publications/tsa.pdf (accessed 2 December 2014).

ROK, 2006. 'National Strategy for Sustainable Development of the Republic of Korea 2006–2010'. Summarised Version. Online, available at: www.rrcap.ait.asia/nsds/pub/korean_nssd_2006-2010.pdf (accessed 2 December 2014).

Sen A., 1999. 'Development as Freedom'. New Delhi: Oxford University Press.

Sustainable Development Knowledge Platform, no date. 'New Growth Strategy: Blueprint for Revitalizing Japan'. Online, available at: http://sustainabledevelopment.un.org/index.php?page=view&type=99&nr=65&menu=1449 (accessed 17 November 2014).

OECD, 2010. 'OECD Environmental Performance Reviews'. Japan, Paris.

UN, 1987. 'Our Common Future. Report of the World Commission on Environment and Development'.

UNCSD, 2012. 'The Future We Want'. United Nations. Rio de Janeiro. Online, available at: www.uncsd2012.org/content/documents/727The%20Future%20We%20Want%2019%20June%201230pm.pdf (accessed 7 October 2014).

UNEP/MAP. 2011. 'Assessment on the Implementation of the Mediterranean Strategy for Sustainable Development (MSSD) for the Period 2005–2010'. Athens. Online, available at: www.odi.org/sites/odi.org.uk/files/odi-assets/publications-opinion-files/7166.pdf (accessed 7 October 2014).

UNEP, 2014. 'Green Economy'. Online, available at: www.unep.org/greeneconomy/aboutgei/whatisgei/tabid/29784/default.aspx (accessed 5 October 2014).

UNESCO, 2000. 'Dakar Framework for Action, Education for All: Meeting Our Collective Commitments'. France: World Education Forum.

UN Habitat, 2013. 'Global Activities Report 2013: Our Presence and Partnership'. Online, available at: http://unhabitat.org/un-habitat-global-activities-report-2013-our-presence-and-partnerships/ (accessed 8 April 2015).

Uras, A. 2014. 'The Review of the Mediterranean Strategy for Sustainable Development'. ESDN Conference 2014. 6–7 November 2014, Rome, Italy.

Weiss, J.D. and Kougentakis, A., 2009. 'Big Oil Misers'. Center for American Progress, 31 March. Online, available at: www.Americanprogress.org/issues/green/news/2009/03/31/5736/big-oil-misers/ (accessed 18 December 2014).

World Bank, 2012. 'China 2030'. Online, available at: www.worldbank.org/content/dam/Worldbank/document/China-2030-complete.pdf (accessed 21 November 2014).

World Bank, 2014. 'Getting to Inclusive Green Growth'. Online, available at: web.worldbank.org/wbsite/external/topics/extsdnet/0,,contentMDK:23184559~pagePK:64885161~piPK:64884432~theSitePK:5929282,00.html (accessed 6 October 2014).

5 Green policies

The concept

The overview provided in previous chapters shows that green jobs, green economy and green growth are interrelated concepts and have comparable goals. They are considered as tools for achieving sustainable development. Thus, as they are used for the same purpose (enabling sustainable development) the same policy tools might be used (regulatory and market based rules, e.g. tax and government spending, labour market regulations).

Policy measures enabling sustainable development have been discussed at length since the Rio 1992 conference. For example, the Rio Declaration included principles promoting the internalisation of environmental costs and the use of economic instruments (Principle 16) as well as eliminating unsustainable consumption and production (Principle 8).

Agenda 21 further elaborated on these principles and called for the development of national strategies for sustainable development[1] incorporating measures for integrating environment and development, providing effective legal and regulatory frameworks, making effective use of economic instruments and market and other incentives, and establishing systems for integrated environmental and economic accounting (Agenda 21, chapter 8).

Some ten years later, the JPoI also identified the need to change the way societies produce and consume, and called for the development of a ten-year framework of programmes for sustainable consumption and production.

In recent negotiations under the UNFCCC it has been agreed that low emission development strategies (LEDs) that adopt appropriate policy measures for low carbon development are indispensable for achieving sustainable development.

The concept of green policy should contribute to sustainable development, as it is based on the double dividend hypothesis. Green policy protects the environment and brings economic benefits.

The *environmental component* of policy is traditionally considered as cost-creating, i.e. as obstacles to production and industries, which, potentially increase unemployment. Environmental considerations are usually dealt with by the regulation of the activities that cause them (prohibition, or at least limiting the behaviour that leads to the deterioration of the environment), or market based instruments (pollution tax, cap and trade).

The regulation approach was used in the first major wave of environmental legislation in the early 1970s. It issues specific instructions in the form of regulations (e.g. as car emissions standards). The direct regulation of activities that cause pollution does not offer any scope for flexibility, and thus can slow down economic growth, e.g. the reduction of GHG emissions based on climate change policies is expected to lower output (between 1 and 3 per cent of gross world product).

A market-based approach gives an incentive, via prices (e.g. pollution tax, a system of tradable emissions permits, a cap and trade system),[2] to limit environmental degradation. Business usually reacts better to market demand than to government regulation. There is wide agreement among environmental economists that a market-based programme that limits carbon emissions by putting a price on them can achieve sizeable results at modest cost. However, there is vivid debate on how quickly market based programmes should be implemented – immediately or gradually.

By the introduction of a pollution tax, the price of pollution is defined, and polluters know what price they will have to pay, but the government does not know how much pollution (and related revenue for the government) they will generate. The tax can be applied domestically and for international trade. Provided that pollution tax on import is comparable to domestic tax (or cost of licenses), it would be acceptable under international-trading rules.

The cap and trade system limits the amount of pollution. Polluters do not know what the price of emissions will be. For the government, if licences are auctioned, the revenue impact is not known. If licences are distributed for free, the potential revenue goes to industry instead of the government.

On the other hand, there is criticism claiming that market based approaches under the concept of green economy are wrong, because environmental and climate crises are not the result of market failure. Thus, it is considered that they cannot be corrected by putting a price on nature.

With regard to the *growth component*, there is no consensus on policies that support growth. The Washington Consensus, post-Washington Consensus and initiatives on green inclusive growth have tried to list policies necessary to foster growth (see Table 5.1).

The Washington Consensus outlines a standard reform package consisting of a set of ten economic policy prescriptions (see Table 5.1) supported by the IMF, the World Bank and the US Treasury Department.

Criticism of the Washington Consensus relates to the measures that are included (e.g. openness to global market) and those that are missing (e.g. institution building), and has widely spread since the 1990s. The World Bank, identified several success factors of East Asian growth (the distribution of income and assets, mass education and state guidance of investment), which was the basis for the promotion of the new institutional economics and a post-Washington Consensus approach. This approach acknowledges that the development process is linked with social relations, the distribution of property rights, work patterns,

Table 5.1 From the Washington Consensus to inclusive growth

Original Washington Consensus	*Post-Washington Consensus*	*Inclusive growth*
Secure property rights	Anti-corruption	Competitive environment
Deregulation	Corporate governance	Government commitment to growth
Fiscal discipline	Independent central bank and IT	'Good policies'
Tax reform	Financial codes and standards	Public sector investment
Privatisation	Flexible labour markets	Labour market deregulation
Reorientation of public expenditures	WTO agreements	Employment and productivity growth
Financial liberalisation	'Prudent' capital account opening	International integration
Trade liberalisation	Non-intermediate exchange rate regimes	Exchange rate management
Openness to FDI	Social safety nets	'Prudent' capital account opening
Unified and competitive exchange rate	Targeted poverty reduction	Social safety nets

Source: Saad-Filho, 2010: 9.

urbanisation, family structures, etc. and offers a rationale for discretionary intervention across a much wider range of economic and social policy than the Washington 9 Consensus (see Table 5.1).

The concept of inclusive growth, that refers both to the pace and pattern of growth, developed in the early twenty-first century. It stresses the importance of growth for poverty reduction and admits that a wide range of policy combinations can deliver specific constraints to sustained and inclusive growth. The Worlds Bank's concept of green inclusive growth is based on the concept of inclusive growth (World Bank, 2012). There are six basic types of interventions that should be considered:

1 pricing and fiscal policies: taxes, subsidies, or subsidy removal;
2 institutions, norms and regulations, and behaviour-based policies;
3 innovation and industrial policies;
4 education and labour markets policies;
5 natural capital, agriculture, and ecosystem services management;
6 infrastructure, building, urbanism, transport, and energy.[3]

Pricing and fiscal policies: taxes, subsidies, or subsidy removal can be used to guide economic behaviour and create environmental and economic benefits. Restructuring of environment-related taxes in a more environmentally friendly way may include:

- the review and development of the system of road fuel and motor vehicle taxes, with a view to promoting more sustainable modes of transport;
- a reduction of subsidies that have negative environmental implications (e.g. reallocating resources from fuel subsidies to spending on education, health, and infrastructure, which helps reaching environmental objectives and increasing economic growth);
- imposing energy taxes.

In OECD countries, environmentally related tax revenues consist mainly of taxes on energy use and vehicles. In the G7 European countries, taxes account for 50–60 per cent of the diesel price and 62–68 per cent of the gasoline price. The OECD recommends raising the consumption tax rate and broadening the base of direct taxes by reducing allowances and deductions (OECD, 2013). Broadening the use of indirect taxes on the consumption of goods and services that are potentially harmful to the environment, e.g. through a carbon tax, would generate revenues that can help the government with fiscal consolidation and/or be used to partly reduce taxes on households and businesses, thereby promoting economic growth. The regressive nature of such taxes should be addressed through ad hoc social benefit schemes. The introduction of other taxes, such as on air and water pollutants, could also be considered.

Political economy considerations play an important part in determining the feasibility of a realignment of fiscal policies with green growth objectives.

A phased approach, supported by communication and complementary policies that reallocate resources, can help build support for reforms. In some cases, resources may need to be allocated temporarily to compensate those negatively affected, even if they are not the poor or needy.

Institutions, norms and regulations, and behaviour-based policies: economic incentives can be complemented with other types of instruments (e.g. building codes for energy efficiency and financial instruments for investments). Policy makers must consider how environmental policies affect businesses and individuals, taking into account their decision making biases and the non-economic incentives that affect behaviours. Also, the impact of changes in technology and lifestyle (e.g. the impact of information/telecommunication technology, increased recreation time, retirement) on environment and nature, taking into account related changes in patterns of settlement, transport, production and consumption, should be considered.

Innovation and industrial policies: green industrial policies need to address both the greening of industries and the creation of new green manufacturing and services, as well as developing and disseminating new technologies.

This process can be eased with specific policies that target:

- the development and dissemination of technologies and innovations, by correcting the effect of a knowledge spillover; and
- the development of new industries and sectors, by correcting the effect of non-environmental market failures (such as coordination failures and capital market imperfections).

Support must carefully balance market failures and government failures given the risks of policy capture and rent-seeking, especially where institutions and civil society are weak.

Education and labour markets policies: green transitions are likely to involve a structural change away from some industries and toward new ones. Policies that facilitate the movement of workers from one sector to another can accelerate the transition and reduce adjustment costs. Where such movement is impeded by skill issues, training programmes can help. These issues may concern the skills necessary for the transformation of existing jobs into greener ones, and those necessary for new jobs. The current skills profile of green jobs will need to change as green jobs change. Skills (and the relevant technology) will be key to delivering environment objectives. However, as there is no single definition of green jobs, the present data on green skills do not provide a sufficient basis for analysis.

As a number of sectors already face shortages regarding the skills needed for green jobs, there is a need to improve the data (Slingenberg *et al.*, 2008). Based on the analysis of the environmental skills programmes in six European countries (the United Kingdom, the Netherlands, Italy, Germany, Bulgaria and Poland), Ecorys identifies that there is considerable variety among EU member states in environmental skills programmes. Usually, firms develop green skills through in-house training to their staff in response to a business need (Ecorys, 2010). There is an even mix of courses that cater for either the high-, medium- or low-skilled – most often financed by a mix of public and private funds. Considerable public funding is provided in respect of the implementation of skills programmes in partnership with companies. There is the potential to promote an exchange of best practice, not only between member states but also between businesses and others involved in the development of green workforce skills.

Any future skills will also depend on demographic trends. In Europe, ageing causes changes in patterns of energy use and consumption, mobility, household location/second homes. However, existing studies (COWI, 2008) show that the ageing of the population does not lead to significant environmental changes. Two major impacts have been identified: the elderly generations' impact on the environment is lower in comparison to the other age cohorts due to their lower mobility, but higher due to higher energy use per person.

Natural capital, agriculture, and ecosystem services management: an excellent way of greening agricultural production is through conservation agriculture, which simultaneously yields environmental benefits (by reducing the pollution of waterways from nutrients, improving soil quality); increases the efficiency of production (by reducing the use of energy inputs); increases resilience (by frequently rotating crops); and increases agricultural productivity in the long run (by reducing erosion and enhancing soil structure).

But for this to work, there needs to be better information underlying the decision making, and better access to this information. For example, greater access to weather and climate information services for farmers can improve resilience in the agricultural system and the overall value chain, including production, post-harvesting, storage and market access. It can also help innovations to succeed (such as in weather-based risk products).

Infrastructure, building, urbanism, transport, and energy: green sectoral interventions can help increase factors of production, push out the production frontier, enhance efficiency, improve resilience, create jobs, and reduce poverty. In some countries, urban congestion and the lack of efficient transportation reduce well-being and hold back economic growth, on top of causing negative environmental effects. Investments in public transit and changes in land-use plans to favour a more compact urban area could reduce air pollution and spur growth (thanks to the benefits from urbanisation and concentration). Multiple benefits can also be reaped from multipurpose infrastructure such as water reservoirs that produce hydroelectricity, mitigate floods, and ensure minimum river flow during drought. And regional integration in infrastructure design and investments can improve the efficiency of the system, for instance by increasing the reliability of electricity generation and allowing for a greater penetration of renewable energy.

Policy choice depends on the priorities, political economy and implementing capacities. This relates to institutional and financial capacity, both. Namely, some interventions create revenue for the governments (e.g. taxes), while implementing others (e.g. training programmes) requires financing. Also, given the high degree of interdependence between social, economic and environmental aspects, every green policy implemented in one sector is likely to impact (either positively or negatively) on other sectors.

For this reason, an integrated cross-sectoral impact analysis of green policies should be carried out in order to provide a coherent evaluation of synergies, side effects and ancillary benefits. An analysis of advantages and disadvantages is necessary to identify the winners and losers, concerning both the required investments and benefits (avoided costs, or added benefits).

This kind of analysis requires the establishment of the baseline, and an estimation of the cost of inaction.

For example, the adoption of energy efficient technology requiring upfront investments (capital expenditure) will reduce energy consumption and expenditure (avoided cost), while possibly creating new jobs and income (added benefit). Added benefits and avoided costs may change depending on the problem analysed, and can be compared with indicators of the historical and current performance of the sector to assess whether the investment can be sustained, and how the economic burden can be allocated across the main actors impacted by the intervention (e.g. public versus private investment).

An analysis of advantages and disadvantages includes four key steps:

1 The establishment of the baseline and an estimate of the cost of inaction.
2 The quantification of the costs of policy intervention.
3 The quantification of the advantages or benefits of policy intervention (estimation of policy impacts across sectors, the evaluation of the direct economic, environmental and social benefits and the potential side effects of the interventions implemented). Integrated simulation models should be used to project the future impact of the interventions on key sectors and indicators.
4 The estimation of the advantages and disadvantages of the policy options, providing basis for informed decision making.

Here we focus on the following channels through which green policies operate: prices and fiscal systems, institutions, investments and behaviour; and innovation and technologies.

The establishment of the baseline and estimate of cost of inaction: this includes an analysis of trends, especially of baseline indicators and the estimation of economic, biophysical, social and cultural damage resulting from inaction (for instance, biodiversity loss and related cost expressed as lost ecotourism GDP).

The quantification of the costs of policy intervention: each policy instrument may imply several typologies of costs, for various stakeholders (e.g. planning, capacity-building, research, operation and management). Some of these costs may be prolonged or even become regular (e.g. operation and management).

The quantification of the advantages or benefits of policy intervention: these benefits include avoided damages, direct and indirect economic and/or biophysical positive impacts, qualitative and/or quantitative social improvements and cultural impacts, and more. The methodologies for quantification can include the estimation of expected returns on investments, cost–benefit and multi-criteria analyses, and are sometimes sector specific (e.g. market and non-market valuation of biodiversity).

Several methods can be used to *estimate the advantages and disadvantages of the policy options*, depending on the information available. These methods include the use of simulation models for ex ante analysis, cost–benefit analysis or a cost-effectiveness analysis, and/or multi-criteria analyses could be applied whereby policy options are ranked against a variety of criteria. Such kinds of analyses enable the forecasting of policy impacts across sectors, including social and cultural ones, and thus requires cross-sectoral expertise and a multi-stakeholder approach.

For instance, the removal of subsidies has three main direct consequences:

- reduction in public expenditure;
- increase of market prices;
- decrease in consumption.

Once the various policy options have been analysed, and further fine-tuned based on the results of the assessment, the best combination of interventions needs to be chosen by policy makers considering the distribution of advantages and disadvantages across key actors in the economy.

The selection of policy options should be based on three criteria,

- the equitable sharing of costs;
- the effectiveness in addressing the issue;
- the promotion of cross-sectoral double and triple dividend opportunities (OECD, 2011; World Bank, 2012).

There is a wide range of *implementing instruments* that be used in combination to stimulate resource efficiency and better environmental performance that could ultimately lead to a better environmental quality. These instruments can be categorised into three:

1. regulatory instruments that mandate specific behaviour;
2. market-based instruments that act as incentives for particular activities;
3. information-based instruments that seek to change behaviour through the provision of information.

Different policy instruments are usually not used in isolation from each other. They typically operate in conjunction.

Regulatory instruments: regulation has been the dominant strategy for pollution control in the EU and across most OECD countries. It can specify an environmental goal (such as the reduction of NOx emissions by a specified date) or mandate the use of a particular technology or process.[4]

Market-based instruments seek to address the market failure of 'environmental externalities' by incorporating the external cost of a firm's polluting activities within the firm's private cost (for example, through taxation), or by creating property rights and facilitating the establishment of a proxy market (for example, by using tradable pollution permits). Market based instruments can be more efficient economically than regulatory standards in achieving a desired reduction in pollution because they act as *incentives* for the development of more cost-effective pollution control and prevention technologies and provide greater flexibility in the choice of technology or prevention strategy. In addition, they may provide the government with a source of revenue that may be used to support environmental and/or social initiatives.

Taxes, charges and fees are among the most commonly used market based instruments. Their successful implementation requires a system of monitoring, revenue collection, and enforcement. Other market based instruments include liability rules, subsidies, low interest loans, direct grants or preferential tax treatment and emission trading schemes.

Information-based instruments are used as an alternative to direct control and to providing incentives for pollution reduction. Examples include requiring public disclosure of environmental information, eco-labelling or environmental product declaration and promoting R&D and demonstration programmes.

Innovation and technology: policy formulation should also consider the implications it has on technical change and innovation. Since innovation is viewed as key to solving many environmental challenges, it is important to understand the mechanisms by which innovation is fostered or limited. Demand-side instruments, such as public procurement, can help foster markets for new products and services, for example through demonstration effects, and counter gaps in the supply of finance at the early stages (OECD, 2011). On the supply side, access to finance is one of the key constraints for business-led innovation, which is inherently risky and may only pay off in the long-term.

Constraints to innovation are not unlike those for green growth in general. They require both market and, at times, non-market responses that encourage risk-taking that may lead to significant rewards. For example, technological barriers could be due to a lack of knowledge that greater international cooperation would solve (some examples are listed in Table 5.3).

National approaches

There are various ways to present national green policies – by broad sector of the economy, by policy approach (e.g. taxation, expenditure, regulation), or by type of sponsoring institution. The presented policies are not exhaustive or comprehensive. They provide examples that reflect the diversity of (successful) approaches.

European Union and its member states

European green policies support the implementation of the European Union's growth strategy, Europe 2020. The main drivers of the growth are addressed by seven flagship initiatives:

1 an industrial policy for the globalisation era;
2 a digital agenda for Europe;
3 Innovation Union;
4 youth on the move;
5 an agenda for new skills and jobs;
6 European platform against poverty; and
7 resource-efficient Europe.

The implementation of policies in these areas should enable the achievement of headline targets that relate to employment, R&D, climate change and energy sustainability, education and fighting poverty and social exclusion (see Table 5.2).

The achievement of these targets requires, inter alia, the implementation of green policies and reversing of unsustainable trends (such as biodiversity loss). The shift to 'a greener economy', i.e. a safe and sustainable low carbon and low-input economy, will require a more efficient and transparent governance structure and adequate investment.

The EC considers that 'green policy does not automatically generate jobs, or destroy them. It's a question of how we do it' (Eckert, 2014). As regards green jobs, there are two main challenges. The first is how to create the right conditions for more new jobs in the wake of the economic crisis; second is the fear that strict environmental obligations could jeopardise jobs and competitiveness.

Table 5.2 EU 2020 targets

Indicator	*2012*	*2020 target*
Tertiary educational attainment age group 30–34 (%)	35.9	40.0
Primary energy consumption (million TOE (tonnes of oil equivalent))	1,584	1,483
People at risk of poverty or social exclusion (1,000 persons)	124,523	Less than 98,000
Gross domestic expenditure on R&D (% of GDP)	2.01	3.00
Greenhouse gas emissions (index 1990=100)	82.14	80
Employment rate total (age group 20–64)	68.4	75.0
Early leavers from education and training (%)	12.7	10.0

Source: Eurostat, 2014 and authors' calculations.

The targeting and coordination of labour market policies and measures are seen as key to enable a transition to a green economy and progress towards Europe 2020 employment objectives. They require fostering skills, promoting mobility and supporting industrial restructuring.

Green policies support the shift towards a resource efficient and low carbon economy, while strengthening the EU's competitiveness. Thus, green policies include energy policies (primarily linked to the production of energy from renewable resources), energy efficiency, waste and water management, air quality, restoring and preserving biodiversity and developing green infrastructure.[5] On a national level, the use of environmental taxation is promoted among policy instruments aimed at achieving environmental benefits and public finance stability. Shifting the tax burden on environmentally harmful activities is considered as a measure that opens the space to lowering the tax burden on growth enhancing categories (e.g. labour) and boosts green growth, green technologies and green jobs (Hogg *et al.*, 2014). NRPs should define short-term measures (for a three-year period) enabling economic recovery and reaching SDGs, coherent with NSDSs.

The *United Kingdom* established the Green Economy Council in January 2011 to support the government on the transition to a green, low carbon economy. It brings together business leaders from a cross-section of industries and sectors, to advise government on green and green growth policies related to business, innovation and skills (such as infrastructure, investment), energy and climate change and environment, food and rural affairs. It will also help to reduce unnecessary regulation.

There are different green taxes, reliefs and schemes for businesses: climate change levy (CCL), the carbon reduction commitment energy efficiency scheme, emission trading, capital allowances on energy-efficient items, landfill tax and aggregate levies (this is a tax on sand, gravel and rock that is dug from the ground or dredged from the sea in UK waters).

The government recognises that green policies (i.e. policies aimed at reducing carbon emissions) are going to impact upon the bills faced by industrial consumers, especially in energy intensive industries. In order to minimise the risk of carbon leakage, in 2013 the government introduced measures to reduce the impact of policy on the costs of electricity for the most electricity-intensive industries, worth around GBP250 million over the spending review period.[6] In December 2013, the government published an update to the Energy Efficiency Strategy.

The United Kingdom is supporting green investment through the GIB. With committed government funding of GBP3.8 billion by April 2015, the GIB is investing in innovative, environmentally friendly areas for which there is currently a lack of sufficient support from private markets.

Since starting operations, the bank has made 22 transactions totalling GBP734 million, which has leveraged GBP2.1 billion total funding from private infrastructure investors (UK NRP, 2014).

Denmark has a long tradition for environmental regulation and protection and also for high green taxes. It introduced a CO_2 tax on fuels in 1992 and engaged in a general reform of its tax system, reducing taxation on income and labour, while increasing fiscal incentives to protect the environment. Denmark's environmentally related tax revenues are 3.95 per cent of GDP (down from 4.1 per cent of GDP in 1994). Energy taxes generate more revenues than other sectors (OECD, 2013). Denmark also taxes scarce natural resources like drinking water: water, energy and urban planning policy are considered as most relevant for greening the Danish economy (Auken, 2011).

The aim of the water policy is to reduce water use and the discharge of waste water in the environment. In parallel with economic growth, Denmark has managed to reduce its water consumption in households and industry by more than 30 per cent in the period 1989–2012, and implemented advanced waste water treatment. This was achieved by environmental targets (that encouraged the development of new technologies that increase the effectiveness and efficiency of water extraction and waste water treatment), information campaigns and market based instruments that have been gradually implemented. Prices on water consumption fully cover the costs of water services.

Since the oil crisis of 1973, Denmark has followed a strategy of both increasing its energy efficiency and diversifying the sources of its energy supply, with an increasing focus on renewables. As a result, the Danish economy is one of the least energy intensive in the world, and CO_2 emissions are decoupled from GDP growth in an absolute sense. Renewables account for about 22 per cent of total energy consumption and the current aim is to increase this to 30 per cent by 2020. A mix of policies has been used to achieve these results. Denmark has some of the highest taxes on energy consumption in order to reduce total energy consumption. Most power plants also deliver district heating, which makes the Danish energy sector very energy efficient. Renewable energy is supported through feed-in tariffs and subsidies during the development, testing and demonstration phases. On the other hand, tax subsidies and favourable price setting for wind-produced electricity and solar panels have also acted as important incentives.

As a result of these policies, Denmark is at the forefront of wind power technology and several other renewable energy technologies. Rather strict regulations and taxation policy have gradually created the demand for the relevant skills and technologies.

Division into urban zones, summer cottage areas and rural zones, developed in 1970, enables the protection of landscapes and agricultural land and prevents uncontrolled development and installations in the countryside. Furthermore, appointed coastal zones are controlling development along the coast, and public access to the coast has been ensured through nationwide regulations. These zones have made urban development a balanced choice, where urban, natural and rural interests in land use are all included.

A number of government initiatives directly stimulate green production and green jobs. In the period 1997–2001 a special public fund labelled 'Green jobs in Denmark' was active. It supported a wide range of activities such as improved energy management in firms, establishment of new environmentally friendly firms and products, nature preservation, waste handling and production of organic food.

As a result, Danish labour has relevant qualifications for a wide variety of 'green activities' spanning from sustainable energy (windmills) to waste handling and energy saving. In many of these areas, Denmark has had the advantage of being an early mover in green technologies. According to the Eco-Innovation index, Denmark is among top five EU countries for eco innovation.[7]

In other areas in research and innovation, recent initiatives include 'Blue jobs via green solutions', Water-efficient industrial production, Innovatorium for building renovation, and Intelligent, sustainable and efficient plant production.

Denmark's vocational training system does not include special training programmes targeted at the green economy. Many of the vocational training programmes include skills that are relevant for 'green technology', also due to the long-lasting influence from the demand side.[8]

There is a potential for the development of new 'green educations' such as, for instance, an 'energy-smith' who will service and maintain green energy suppliers in private homes as well as companies, optimise installations for electric cars, etc. Such new job functions could be related not only to the manufacturing sector, but just as much to the service and installation sector targeted at private households or companies. However, the economic crisis has led to massive layoffs in the construction sector and in manufacturing, leaving many skilled workers unemployed. As a result the challenge is not the lack labour, but the lack of jobs to employ the existing qualified workforce. Conversely, the economic crisis has stimulated the interest for export markets and thus shifted the focus of the debate on 'economy greening' from the issue of environmental protection to the target of increasing international competitiveness and employment.

Strong environmental policy has stimulated the creation of green production and green jobs. It has created demand from both consumers and firms for products that could reduce environmental damage and the use of the resources subject to green taxes.

As regards informative measures, eco-labelling has been most successful in the Nordic countries where eco-labels are popular with consumers and most large producers have voluntarily followed suit.

The United States of America

The Green Jobs Act of 2007 launched measures related to education and labour market policies and industrial policies.

The Act supported the identification of needed skills and the development of training programmes, and committed the federal government to funding job training (with US$125 million) for 35,000 people a year to work in environmentally friendly fields such as energy efficient buildings and construction, renewable electric power, energy efficient vehicles, and biofuels development. In addition, US$80 million was allocated for 'just transitions' – green job re-training, for example teaching construction workers competency in green construction techniques, US$25 million to create 'green pathways out of poverty', i.e. job training for low-income people in fields like solar power installation and green roofing, and US$20 million for renewable energy and energy efficiency research.

The Green Jobs Act was later (in 2009) included in the Energy Independence and Security Act and was appropriated US$500 million through the American Recovery and Reinvestment Act (ARRA). ARRA funded a national workforce training programme, Pathways out of Poverty, targeting individuals living below or near the poverty level to provide them with the skills needed to enter the green job market, focusing on the energy efficiency and renewable energy industries.

The challenges related to the adoption of green policies can be illustrated by the American Clean Energy and Security Act of 2009 (ACES), also called the Waxman-Markey Bill. It tried to introduce a variant of an emissions trading plan, similar to the European Union Emission Trading Scheme. The bill was approved by the House of Representatives in 2009, but was defeated in the Senate.[9]

The vote by the House of Representative was the first time either the House or Congress had approved a bill meant to curb the heat-trapping gases that scientists have linked to climate change. According to the study realised by the Heritage Foundation, the bill would have little impact on world temperatures, but would by 2035:

- reduce aggregate GDP by US$7.4 trillion;
- destroy 844,000 jobs on average, with peak years seeing unemployment rise by over 1,900,000 jobs;
- raise electricity rates by 90 per cent after adjusting for inflation;
- raise inflation-adjusted gasoline prices by 74 per cent;
- raise residential natural gas prices by 55 per cent;
- raise an average family's annual energy bill by US$1,500; and
- increase inflation-adjusted federal debt by 29 per cent, or US$33,400 additional federal debt per person, again after adjusting for inflation (Beach *et al.*, 2009).

On the other hand, the Congressional Budget Office estimates differ significantly from those of the Heritage foundation, whose results were also challenged by analyses of the Center for Climate and Energy Solutions.[10]

The Congressional Budget Office has concluded that the Waxman-Markey Bill would reduce the projected average annual rate of growth of GDP between 2010 and 2050 by 0.03 to 0.09 percentage points. That is, it would trim average annual growth to 2.31 per cent, at worst, from 2.4 per cent. Over all, the Budget Office concludes, strong climate change policy would leave the American economy between 1.1 per cent and 3.4 per cent smaller in 2050 than it would be otherwise.

The main policies are considered 'clean', and not necessarily 'green'. This relates primarily to decarbonising the energy sector (through the promotion of carbon capture and sequestration technologies, as well as adopting federal GHG emission standards for new and existing power plants), energy efficiency and developing clean fuels (primarily biofuels, fuel efficiency). This is also seen as a way to combat the climate change. Renewable tax credits support the use of renewables. Several states have emission trading schemes and GHG goals (Californian emission trading system and the goal to reduce GHG levels to 1990 levels by 2020, and to further reduce by 80 per cent of that by 2050).

As regards environmental protection, policies focus on restoring ecosystems and reducing mercury pollution from power plants.

Canada

The main challenge related to identifying and estimating the effectiveness of the green Canadian policies is a lack of definition of green economy, and knowledge gaps related to policies that promote EGS, as opposed to policies that provide for greening the whole economy. This gap is particularly acute on the question of which Canadian green policy priorities make sense, given Canada's national circumstances.[11]

Without a definition and indicators, it is difficult to properly plan and undertake initiatives that contribute to the transformation to a green economy. Moreover, this limited definition creates a false perspective on the role that other sectors of the economy currently play in the green economy, and limits overall understanding of how the economy as a whole can move onto a more sustainable path.

One such sector is the primary extraction of natural resources. While this sector is not green, the green policies should also deal with the most important challenges and opportunities of the economy. In the case of Canada, the relevant challenges are macro-economic (such as export diversification, changing demographics, a high value currency, unemployment and high government and consumer debt).

There are economic and environmental concerns that Canadian exports are too heavily weighted in primary resources. In order to foster the development of export-oriented sectors, the development of technology and new green industry can alleviate current challenges.

China

China's twelfth five-year plan 2011–2015 includes a section on 'Green Development'. 'Green Development' identifies six strategic pillars: climate change, resource saving and management, circular economy, environmental protection, ecosystem protection and recovery, water conservation and natural disaster prevention. These pillars entail several new binding targets (e.g. carbon emissions per unit of GDP to be reduced by 17 per cent by 2015, NOx and nitrogen air emissions to be reduced by 10 per cent by 2015) and detailed policy guidelines, primarily related to energy savings, emission reduction and industry.

The energy saving policy emphasises the reduction of energy waste and the increase of energy efficiency. It relies on market instruments (taxes and fees, subsidies and other instruments), international cooperation and resource control. Clean energy, energy efficiency and clean vehicles are identified as growth sectors. Low-polluting vehicles are tax deductible, and a zero tax rate applies for fixed asset investment in cogeneration projects. Energy prices are used to reduce energy consumption. International cooperation promotes technological development (e.g. for clean coal capture and sequestration with the United States, electric vehicle programme with Germany).

The emission reduction policy focuses on the reduction of pollution, such as CO_2, sulphide, oxynitride, particulate matter and polluted water. It relies on market instruments (taxes, fees and subsidies), information measures and resource controls, e.g. purchase of new, less air polluting cars is subsidised, there are special emission criteria for heavy diesel cars. Seven city/regional emissions trading schemes have been introduced. Industrial policies are expected to guide the development of green industry.

Japan

There are two main sectors in which Japan's green policies are recognised as successful practices: agriculture and energy (World Resources Institute, 2011).

Relevant agricultural policies are behaviour based, and enable decent jobs and sustainable production, such as participating in the Fairtrade Labelling Organizations International.[12]

The fair trade label guarantees that the farmer, acting through a fair trade farmer cooperative, has been paid a fair trade price. Such price is based on the average cost of sustainable production (as opposed to the volatile global commodity market prices). Furthermore, the label ensures that farm workers are protected by appropriate labour standards and that a fair trade 'premium' (in addition to the fair trade price) has been paid to contribute to community projects, i.e. participation in the Fairtrade Labelling Organizations International provides for greening agricultural production, which is not necessary situated in the country that supports the programme.[13]

The World Resource Institute identified the green stimulus package and 'Top Runner programme' as successful green policy programmes in the energy sector.

Japan adopted the green stimulus package in 2009 (US$154 million) to foster environmentally friendly technologies; implementing measures including subsidies and tax breaks to provide cheaper solar power to households and support energy efficient equipment and vehicles. The policy included tax breaks for consumers (up to US$2,500) for purchases of 'green' cars, as well as subsidies of 5 per cent for the purchase of energy efficient televisions and other appliances.

The Top Runner programme enabled defining energy efficiency standards for 21 products (e.g. TVs, air conditioners, vending machines) sold in Japan. On a regular basis, all the products currently available in a category are tested to determine the most efficient model. That model's level of efficiency is defined as the new baseline. Manufacturers have the obligation to make efforts to achieve the new baseline within four to eight years. If a manufacturer does not meet the target or fails to make a good faith effort, this fact is publicised. Products that meet the energy efficiency standard receive a Top Runner label; products that do not meet the baseline are labelled differently. This drives other companies to try to make even more efficient models to compete. This programme provides for continual improvements of energy efficiency standards.

Japan's approach to green policies is characterised by a strong emphasis on performance standards and negotiated agreements with industry. Industry and consumers are encouraged to purchase environmentally friendly products by providing various forms of subsidies rather than by including environmental costs in the price of goods and services.

These approaches have helped Japan to become a leader in a range of environment- and climate-related technologies and to improve the environmental performance of manufacturing processes and products. They promote incremental technological changes.

According to the OECD, more cost-effective policy instruments, particularly market-based instruments that apply to the economy as a whole (such as carbon tax, cap and trade system) and not just to particular sectors, would provide better incentives for achieving environmental objectives and for promoting eco-innovation (OECD, 2010).

Republic of Korea

By adopting three key documents in 2008, Korea defined a framework for the development of green policies.[14] They had three primary objectives:

- the promotion of 'eco-friendly new growth engines' for the national economy,
- the enhancement of South Korea's quality of life; and
- a contribution to international efforts to fight climate change.

Target spending of 2 per cent of the nation's GDP on green growth was also defined, thus the programme was domestically known as the 'Green New Deal'.

In the first phase, the policies (and related funding) focused on infrastructure, with spending shifting toward export-focused green-tech R&D over time.

Within the period 2009–2013, ten green policy categories were defined:

1. effective mitigation of GHG emissions;
2. reduction of fossil fuel use and enhancement of energy independence;
3. strengthening the nation's capacity to adapt to climate change;
4. development of green technologies;
5. the 'greening' of existing industry and promotion of green industries;
6. advancement of the industrial sector;
7. engineering a structural basis for the new green economy;

8 greening the land and water and building the green transportation infrastructure;
9 bringing the green revolution to the daily lives of the nation's citizens; and
10 becoming a role model for the international community as a green-growth leader.

Under the first of these policy categories, specific activities included:

- building a national GHG inventory reporting system and emissions trading mechanism;
- developing cost-effective mitigation strategies for buildings, transportation and industry; and
- pursuing forestation and sustainable forestry management programmes to increase carbon absorption.

Similarly, under the seventh policy category (greening the land and water and building the green transportation infrastructure) planned activities included the gradual introduction of a GHG emissions trading system with an initial trading volume of KRW500 billion (US$0.4 billion) in 2013 and KRW2 trillion (US$1.6 billion) by 2020; providing public credit guarantees to the green-tech and green-industry sectors, with investment increasing from KRW2.5 trillion (US$2 billion) in 2009 to KRW7 trillion (US$5.6 billion) in 2013 and KRW8 trillion (US$6.4 billion) by 2020; and overhauling the nation's taxation system to encourage GHG emissions reduction, energy efficiency and other 'green' economic activities.

On a sectoral level, green policies were implemented through stimulus package and by fostering energy efficiency.

The ROK arranged a US$30.7 billion stimulus package in the energy sector. It supported the government's efforts in various environmental projects including the development of renewable energy resources, energy efficient buildings, low carbon vehicles, expansion of railways, and water and waste management.

The government has also established the mandatory reporting of carbon emissions by all carbon and energy-intensive industries, and a cap on GHG emissions.

The benefits of the green stimulus package include increased employment in greening sectors, improvements in energy and income security and reduction of GHG emissions (by 30 per cent below the nation's business-as-usual levels by 2020). Also, the green policies support the creation of demand, and thus improve market opportunities for green products.

The improvement of energy efficiency is guided by national plans for new and renewable energy (NRE). In 2008 Korea adopted its Third National Basic Plan for NRE R&D and Deployment (Third Basic Plan). It identified NRE development and deployment targets. According to the Third Basic Plan, under a business-as-usual scenario, NRE's share of the primary energy supply in Korea was projected to account for 3.6 per cent by 2015, 4.2 per cent by 2020 and 5.7

per cent by 2030. Under the Third Basic Plan's targets, however, the ROK seeks to increase NRE's overall share to 4.3 per cent by 2015, 6.1 per cent by 2020 and 11 per cent by 2030.

The sector specific and general green policies are coherent – green policies support the implementation of the basic plan. The second green growth policy category (reduction of fossil fuel use and enhancement of energy independence) supports the implementation of the Third Basic Plan's targets for NRE deployment, to be implemented by industrialising NRE production and by applying renewable portfolio standards, obligating utilities to meet increased renewable resource utilisation levels.

Green Korean policies were planned to be implemented nationally and aboard, as it aims to become a global green-energy/green-tech leader. For example, even under the five-year plan's second policy category (reducing fossil fuel use and enhancing energy independence), the implementing measures relate to encouraging business enterprises specialising in overseas resources development to explore cooperative green-growth endeavours abroad.

Research and innovation policies also support greening the Korean economy, but also fostering its role in global markets. Thus, green policies support the development of green technologies (solar cells, high-energy fuel cells, integrated coal gasification combined cycle and smart grid technologies) and the further development of strategic industries (automotive, chemicals, semiconductors and steel).

Specific targets for the period until 2013 were to raise Korea's global market share in green energy technologies to 8 per cent and in the export of green goods (in strategic industries) to 15 per cent (from 10 per cent in 2009).

These governmental measure had an impact on access to finance and companies' business strategies. In 2009, the Korea Development Bank, which already had revealed aspirations to become Asia's leading investor in GHG emissions-reduction projects, announced plans to provide approximately US$50 million in financing for a Malaysian palm oil plantation biomass renewable energy project expected to yield 300,000 certified emission reductions. In 2009, the Korea Trade Centre (KOTRA) and nine Korean companies launched an initiative aimed at identifying investment opportunities in renewable energy-focused clean development mechanisms and project development and investment opportunities, including biomass/biogas, wind, hydropower and energy efficiency projects.

Similarly, the Korean leading industrial conglomerates also have taken up the challenge, with many already pursuing plans to aggressively expand their R&D, manufacturing and export programmes in the areas of wind, solar, electric/hybrid vehicles, smart grid and related green technology.

Korean green policies are promoted as a business model, offering successful examples as to how taxation policy, transportation and energy infrastructure support the development of the economic growth opportunity presented by renewable energy technology development.

In the period 2008–2010, Korean green policies were developing under a low carbon, green growth policy framework. In 2010 the Basic Law on Low Carbon and Green Growth entered into force. It relies on new engines of growth, primarily green financing (guarantees, bank loans and venture capital) and promoting of green innovation and private investments.

Australia

The World Resource Institute in 2011 identified two categories of Australian green policies as examples of successful practice (World Resources Institute, 2011). These fall into following categories:

- education and labour market policies; and
- institutions, norms and regulations, and behaviour-based policies.

The Green Skills Agreement seeks to build the capacity of the VET sector to deliver the skills for sustainability and required in the workplace and to enable individuals, businesses and communities to adjust to and prosper in a sustainable, low carbon economy. This will be achieved by: embedding skills for sustainability practice and teaching in VET, within the requirements of the national regulatory framework; the upskilling of VET instructors and teachers to deliver skills for sustainability; the strategic review of training packages (sets of nationally endorsed standards and qualifications for recognising and assessing people's skills) to embed sustainability knowledge, skills and principles; and implementing a transition strategy to re-skill vulnerable workers.

The Council of Australian Governments endorsed the Green Skills Agreement in 2009 that commits the Australian government and state and territory governments to working with training organisations and businesses for sustainability skills.

The programme enabled the participation of more than 1.7 million Australians in VET each year through a national network of over 4,000 public and private registered training providers. In addition, the programme increased market entry opportunities for green industries, and support for the reduction of GHG emissions through the market transformation of construction, energy, mining, manufacturing, transport, agriculture, and administrative and service-related industries was provided (World Resources Institute, 2011).

Australia changed government in 2013. The new government has fundamentally changed its environmental policy, disbanded the Climate Commission and the Senate and some other institutions (such as the Climate Change Authority and the CEFC) and green policies. A carbon tax was replaced by new measures, including the Emissions Reduction Fund, which provides incentives to businesses for reducing emissions.

Brazil

Brazil uses a significant share of renewables in its energy mix. Use of ethanol is widespread in the transport sector. Thus the energy and transport sectors have relatively limited opportunities for emission reduction and greening the economy. The most significant potential relates to deforestation. Also, Brazil has made significant progress in greening its urban development.

Green policies (mostly low carbon ones) are expected to generate benefits in three ways: by fostering economic growth, achieving long-term environmental sustainability and limiting climate change.

Brazil has made impressive progress in addressing many of its development issues. Forest protection policies and programmes have seen considerable results and deforestation has reduced dramatically. Brazil has also started applying innovative solutions that reconcile growth and sustainability.

Despite this progress, approximately 40 per cent of Brazil's gross carbon emissions still come from deforestation. Together with agriculture and ranching, 75 per cent of Brazilian emissions derive from changes in land use.

There is significant potential to further reduce deforestation. According to the World Bank, the use of various instruments, such as increasing the use of pastures and reintegrating degraded areas to the production cycle, thus avoiding encroachment on new areas, could reduce deforestation by up to 68 per cent by 2030, compared to the reference scenario estimated for that same year (World Bank, 2010). Market based mechanisms are insufficient to take advantage of all the opportunities and to mitigate emissions, thus regulation is necessary.

Public transportation policies in cities could reduce emissions by 26 per cent by 2030. Combining these with an increased use of ethanol could double these reductions. The waste management sector accounts for the lowest share of Brazilian emissions, 4.7 per cent in 2008, but the implementation of adequate policies could reduce sector emissions by up to 80 per cent by 2030.

Urban planning policies thus have a significant role in greening the economy. In this respect, Brazil adopted a policy approach that combines environmental regulations to control polluting industries with financial support for city services, including public transportation, waste management, local infrastructure improvements, and public education. A successful example of this approach is the city Curitiba, located in Parana, Brazil. It has implemented several innovative systems to create jobs, improve public transportation accessibility, promote housing development, and improve waste management. The city has integrated a 'radial linear-branching pattern' to protect density by diverting traffic from the city centre and to protect green areas by encouraging industrial development along radial axes. Curitiba has initiated a Bus Rapid Transit System and established the Curitiba Industrial City on the city's west side, which has strict environmental regulations and does not allow 'polluting' industries.

Good green practices in the energy sector are also developed on a city level. The municipal government of Sao Paulo defined a GHG reduction target (30 per cent of 2005 levels by 2012). The main implementing instruments focused on transportation, renewable energy, energy efficiency, waste management, construction and land use. Sao Paulo was the first city in Brazil to pass such legislation.

The success of the policy is measured by air quality, the amount of GHG emissions and indicators related to public transportation access and use, expected landfill waste reduction and expected reductions in deforestation due to the requirements for certified wood for use in construction.

Mexico

Mexico is using a market based and regulation approach to foster a green economy and balance its economic, social and environmental goals. The Climate Change Act, forest protection legislation, and legislation and funding mechanisms adopted to increase renewable energy and energy efficiency are examples of Mexico's green policies.

Taxation policy is also used to pursue environmental and social objectives. The extended use of environmentally related taxes, and reforming environmentally harmful subsidies is recognised as an opportunity to tackle climate change and other environmental problems. It also serves to broaden the tax base.

As regards subsidies, there is significant potential for the removal of environmentally and socially unjustified subsidies. For instance, prices of transport fuels are regulated via a price smoothing mechanism that results in an implicit subsidy at times of high world oil prices. This subsidy represented a net expenditure of 1.2 per cent of GDP in 2011, despite the fact that the government progressively raised fuel prices in the late 2000s.

Overall, energy subsidies, including those for electricity consumption in the agricultural and residential sectors, averaged about 1.7 per cent of GDP per year over 2005–2009 (OECD, 2013).

Most of these subsidies have benefitted the rich more than the poor. The poorest 20 per cent of the population captures only 11 per cent of residential electricity subsidies and less than 8 per cent of transport fuel subsidies. Similarly, 90 per cent of agricultural price support and 80 per cent of electricity subsidies for water pumping benefit the richest 10 per cent of farmers. In 2008, energy subsidies cost more than twice the amount spent on anti-poverty programmes.

Removing environmentally harmful subsidies and reconciling environmental and poverty alleviation objectives thus represents significant opportunity for achieving environmental and social goals.

The promotion of green activities is expected to increase the demand of some occupations and require further training for others. Thus, the need for linking employment, education and professional training policies with environmental policies is recognised (e.g. the National Climate Change Law and the National Programme for Sustainable Production and Consumption), with the objective of matching labour supply and demand.

Turkey

Policies that enable environmental protection (in line with EU standards), with special emphasis given to resource efficiency, clean production and consumption, and reduced emissions intensity as well as those that foster employment and innovation in environment-related sectors, and are socially inclusive and welfare improving in Turkey are considered green. These policies accompany economic policies, primarily those aimed at increasing the level of investment and its efficiency, generation of jobs and higher labour productivity, as well as other measures to enhance competitiveness and mitigate risks (World Bank, 2008).

Turkish environmental policy instruments are mostly limited to energy taxes, environmental impact assessments, and pollution penalties. In addition, fees are charged for municipal services to cover waste water and solid waste management, and a nominal charge is applied to irrigation water.

It is recognised that additional policy measures, such as additional pollution taxes, emission trading and permits, and abatement investments towards reduced energy intensities are necessary in order to achieve green growth.

The green policy mix includes regulatory, market based and information measures. Turkish Environmental Law now recognises the use of carbon markets for climate action. The establishment of a carbon market is planned for 2015. The policies should address green innovation and green growth constraints alike and deal with the main strategic sectors.

The strategic sectors that have significant greening potential in Turkey face a number of constraints – usually in the form of market, institutional or policy failure. Market failures can arise if the production of a good also creates some form of residual that is not 'priced' in the market. This is often the case with pollution – where the market does not reflect the true social cost of producing goods or services. In this case the rationale for government intervention is to internalise these environmental 'externalities,' which is viewed as correcting a market failure. An example of this is CO_2 or PM_{10} emissions from the cement or iron and steel sectors. According to the 'polluter-pays-principle' the sector should be responsible for absorbing the additional costs of reducing pollution. Institutional failure is more difficult to define precisely, but can arise when there is a lack of government intervention or institutions to regulate certain behaviours. For example, the absence of a mechanism for water pricing in agriculture can be viewed as an institutional failure that leads to inefficient use. Policy failure is usually the circumstance where policy fails to achieve its intended objective. Many energy policies fall into this category, where subsidising fossil fuels runs counter to efficiency objectives – resulting in over-consumption.

Re-incentivising through policies is more complicated than it first appears, however the EU Directives offer some guiding principles that Turkey has to follow. Some examples of the major constraints faced, applicable EU Directives and some possible policy responses to guide the sector in its transition to a greener development path are presented in Table 5.5.

Final remarks

In order to be considered as a green, policy should support sustainable development and have environmental and growth components, both. It can be primarily linked to green jobs, green economy and/or green growth.

Table 5.3 Policies to promote green innovation in Turkey

Policy challenge	*General assessment for Turkey*	*Policy options*
Insufficient demand for green innovation	Government subsidies to SMEs for green innovation exist but no market-based instruments, standards or regulations	– Demand-side policies, such as public procurement, standards and regulations, in specific markets and circumstances – Market-based instruments to price externalities and enhance incentives
Lack of innovation capability	Private-sector-led green innovation is advanced by large, exporting firms	– Broad-based policies to strengthen innovation
Technological roadblocks and lack of radical innovation	Leading innovation concentrated in the academic environment	– Investment in relevant R&D, including thematic and mission-oriented research – International cooperation
Research and investment bias to incumbent technology	Absolute level is low, but can be scaled-up and more strategic	– R&D support, tax incentives – Adoption incentives/subsidies – Technology prizes
Lack of finance	More opportunities need to be created, including PPPs	– Co-investment funds – Market development
Regulatory barriers to new firms	Not a major constraint	– Regulatory reform – Competition policy – Front-runner approaches
Lack of capabilities in SMEs to adopt green innovation	Capacity needs are high (i.e. professional education, vocational programmes)	– Access to finance – Skills development – Linking SMEs tó knowledge networks – Improving information supply – Reducing regulatory burdens
Non-technological innovation	Problematic in major urban areas (e.g. transport logistics)	– City and transport planning – Regulatory reform
International technology transfer	Weak implementation of intellectual property rights (IPR) and patent pools	– Development of capabilities – Trade and investment policies – IPR protection and enforcement – Voluntary patent pools and collaborative mechanisms

Source: OECD, 2011.

Table 5.4 Policy options to green growth constraints

Green growth constraints	*General assessment for Turkey*	*Policy options*
Inadequate (soft) infrastructure	Weak institutional support for entrepreneurial activities	– Public–private partnerships – Public investment – Tariffs – Transfers
Low human and social capital and poor institutional quality	Governance issues exist, but country is a dynamic and emerging economy	– Subsidy reform/removal – Growing and stabilising government revenue
Incomplete property rights, subsidies	Weak implementation	– Review, reform or remove
Regulatory uncertainty	Long planning tradition exists, but checks and balances need strengthening	– Set targets – Create independent governance systems
Information externalities and split incentive	Some information campaigns exist (eco-labelling), but more can be done to scale up and focus in on EU-bound products	– Labelling – Voluntary approaches – Subsidies – Technology and performance standards
Environmental externalities	Prices do not reflect scarcity or other non-market values (e.g. pollution taxes)	– Tradable permits – Subsidies – Taxes
Low returns to R&D	Firm-specific R&D robust (e.g. large, exporters)	– R&D subsidies and tax incentives – Focus on general purpose technologies
Network effects	Exist, but need to be enlarged to take advantage of possible synergies (e.g. industrial parks or even in cities)	– Strengthen competition in network industries – Subsidies or loan guarantees for new network projects
Barriers to competition	A vibrant private sector has developed	– Reform regulation – Reduce government monopoly

Source: OECD, 2011.

A narrow approach considers a policy green if it is related to EGS (i.e. sector specific, environmental policies). A more comprehensive approach deals also with the growth component and macroeconomic policies.

Green policy should support the transition to a green economy over the short-, medium- and long-term, including induced job creation/destruction from the impacts on labour supply and productivity, wages, taxation, energy prices, and the overall price level. For example, increasing taxes on environmental pollution and reducing labour taxes can create green jobs and can be considered as green policy. Such policy should also consider the skills that are required for the green economy: education and training policies, as well as taxation are thus green polices.

Almost all policy actions can be green and contribute to sustainable development (development, social, economic, etc.). It is essential that the policies allow the protection of the environment and bring economic benefit. Thus, green policies include macroeconomic and microeconomic policies alike. They should also enable correcting market failures (by use of market based instruments for environmental externalities and/or regulation).[15]

Green policies should also deal with current and emerging challenges in a national context. As a result, they are extremely diverse, which makes it difficult to specifically target individual measures and monitor implementation.

The examples represent a contribution to the ongoing debate on how to put in practice the 'green growth' concept, the instruments that could be used, and the obstacles that are likely to emerge.

However, while some of the policies are accompanied with impact assessment, quite limited information is available on implementation costs. The effects of green policies depend on the policy measures taken, the methodological approach and country specific circumstances.

The ILO uses a double dividend hypothesis, according to which policy measures can simultaneously achieve economic benefits (employment gains) and environmental improvements (ILO, 2013). Various studies provide different scenarios and approaches, but the majority show that the net gains from green growth in employment are in range of 0.5–2.0 per cent (ILO, 2013). At the global level, it is estimated that imposing CO_2 tax and using the resulting revenues to cut labour taxes could create up to 14 million new jobs (IILS, 2009).

The current policies of most governments do little to address widening income disparities and structural unemployment (which are relevant for the creation of green and decent jobs) that are the most serious problems confronting the global economy (WEF, 2014, cf. Piketty, 2014). The shift in policy from austerity to investments in low carbon economy, improving labour market policies, skills, education and innovation could enable job creation and also support green growth. Other relevant green policies include restrictions of GHG emissions, adaptation to climate change and other environmental policies.

The relative priority put on each of these elements depends on national circumstances (Bouzaher *et al.* 2013), and primarily the availability of various forms of capital, including natural capital.

Table 5.5 Example policy responses among key strategic sectors

Sector	*Greening potential*	*Major constraints*	*Major EU directives and progress to date*	*Possible policy responses*
Agriculture	Promoting soil conservation practices such as zero-tillage and pasture improvement	Lack of knowledge; resistance to change; limited availability of special equipment; and insufficient access to capital to finance the initial investment		Agri-environment schemes such as the environmentally based agricultural land protection programme (ÇATAK); pilots sponsored through priority axis 2 of the IPARD programme
				Local education and promotional campaigns to raise awareness of conservation tillage
	Irrigation water conservation	Inefficient use of water; leakages; nitrate pollution	Nitrate Directive	Public sector support for the diffusion of new knowledge through training and advice (extension) (e.g. support programmes for nutrient reduction in meeting the Nitrate Directive)
				Water pricing reflecting scarcity value coupled with technical assistance programmes on water-saving technologies and drainage (a form of tax)
Automotive	Energy efficiency	Low-cost inputs leading to inefficient use		Full- or social-cost of inputs used (a form of tax)
	Waste – recycling	Product innovation with greater recycled content	End-of-life vehicles	Producer responsibility – with take-back programmes fed into secondary recycling markets; eco-labelling
Construction (buildings)	Energy efficiency	Institutional arrangements and financing		Taxes on emissions/output
Cement	CO_2 emissions	Market-failure (CO_2 not priced)	IPPC/ ETS	Taxes on emissions/output; with eventual participation in emission trading system (ETS)
	Hazardous waste		Hazardous Waste	Outright bans – restrictions; fines or penalties for violations

Electronics	Mercury, cadmium, lead and other toxic substances contained in e-waste	Short life-cycles leading to increasing volume; stakeholder coordination issues with the WEEE Directive*	Hazardous Waste WEEE Directive	Life-cycle regulations requiring safe disposal per the Hazardous Waste and WEEE Directives; creation of hazardous (e-)waste collection points
Iron and steel	CO_2 and PM10 reductions	Blast furnace technology is dated and emissions-intensive	IPPC/ETS	Taxes on emissions/output; with eventual participation in emission trading system (ETS)
	Reduction of emissions of dust, CO, heavy metals, nitrous oxides, organic gases/steams, dioxins and furans	Importation of 'dirty' scrap requiring pre-treatment	Waste and Hazardous Waste Directives	Regulations on imported scrap; greater domestic recycling content to reduce reliance on imported scrap; eco-certified or labelled products
Machine industry	Handling of waste and e-waste	Institutional and market failure (absence of institutional structure for handling waste and the potential of recycling e-waste in secondary markets)	WEEE Directive	Waste management fees (both regular and hazardous); regulations on landfilling; eco-labelling
White goods	Handling of waste and e-waste	Institutional and market failure (absence of institutional structure for handling waste and the potential of recycling e-waste in secondary markets)	WEEE Directive	Waste management fees (both regular and hazardous; regulations on land-filling; eco-labelling

Source: World Bank, 2013: 43-44.

Note
* WEEE is waste electrical and electronic equipment.

A mix of policies, aimed both at producers and consumers enables greening. They should be focused and clearly prioritise key areas. In particular, the scope could concentrate on the following four elements: market-based approaches, regulations and standards, measures to incentivise R&D, and information-based instruments. However, as there is no agreement on optimal environmental policy instrument, nor on growth policies, building on existing policy instruments and policy coherence remains a challenge.

Notes

1 Various approaches to this issue at national level are presented in Chapter 4, Table 4.3.
2 A cap is defined by licences to emit a specified pollutant. A business that creates more pollution than it is licensed to has to buy additional licenses from other parties; a firm that has more licenses than it uses can sell its surplus. This gives incentives to reduce pollution.
3 This is by no means an exhaustive list, e.g. the Rio+20 Conference identified international trade as an engine for development, sustained economic growth and the transition to a greener economy.
4 The IPPC Directive is a typical regulatory instrument. Its implementation can benefit from parallel approaches like negotiated compliance, market-based instruments, and informational tools.
5 There are also other relevant, sector specific green policies, e.g. participation in the Fairtrade Labelling Organizations International is recognised as behaviour based green policy relevant for agriculture. For more about this, see section on Japan.
6 Carbon leakage is the term used to describe the situation that may occur if, for reasons of costs related to climate policies, businesses were to transfer production to other countries that have laxer constraints on GHG emissions.
7 The top ranking EU countries for eco-innovation are Sweden, Finland, Germany, Denmark and the United Kingdom, which score, relative to the EU average (=100), 138, 138, 132, 129 and 122 respectively (Eurostat, 2014).
8 E.g. programmes aimed at the installation of energy-saving equipment in buildings could be included under green heading.
9 For more about the bill and other initiatives see the Center for Climate and Energy Solutions website, online, available at: www.c2es.org/federal/congress/111.
10 For more see the Center for Climate and Energy Solutions website, online, available at: www.c2es.org/federal/congressional-policy-brief-series.
11 This relates to vast land areas, climate and the importance of resource extraction for the economy. See chapter on green growth.
12 This is an initiative that includes Japan, more than 20 other countries in Europe, North America as well as Australia and New Zealand.
13 For example, fair trade is well developed for coffee production, where coffee is shade-grown, which provides songbird habitat and other environmental benefits compared to coffee planted where the forest canopy has been removed. With growing consumer support, fair trade has achieved a significant market share in many product categories in the 70 countries where fair trade products are sold. In some national markets fair trade accounts for 20–50 per cent of the market share in certain products. In 2008, fair trade certified sales amounted to approximately US$4 billion worldwide, benefiting an estimated six million people.
14 The National Strategy for Green Growth (2009–2050), the Five-Year Plan for Green Growth (2009–2013) and the Framework Act for Low Carbon Green Growth. In 2009, South Korea's National Assembly adopted the Basic Law on Low Carbon and

Green Growth, which targets spending 2 per cent of the ROK's GDP annually to stimulate green businesses and projects, encourages programmes to mitigate GHG emissions, and delegates its implementation to several key governmental ministries.

15 Market based instruments are considered effective implementing instruments. On the other hand, there is criticism claiming that postulates promoted under the concept of green economy are wrong and that the current environmental and climate crises are not the result of market failure, thus that they cannot be corrected by putting a price on nature.

References

Auken, I., 2011. *Danish Best Practices in the Context of Green Economy.* Sharing Green Economy Best Practices towards Rio+20. Warsaw, 11 October.

Bouzaher, A., Guadagni, M., Sahin, S., Meisner, C. and Kasek, L.P., 2013. *Turkey Green Growth Policy Paper: Towards a Greener Economy.* Washington: World Bank.

COWI, 2008. *Environment and Ageing.* Kongens Lyngby. Online, available at: http://ec.europa.eu/environment/enveco/others/pdf/ageing.pdf (accessed 16 November 2014).

Eckert, D., 2014. *Green Jobs for Inclusive Growth.* Speech. 8 July. Online, available at: http://ec.europa.eu/environment/news/efe/articles/2014/07/article_20140708_01_en.htm (accessed 5 November 2014).

Ecorys, 2010. *Programmes to Promote Environmental Skills.* Online, available at: http://ec.europa.eu/environment/enveco/industry_employment/pdf/environmental_skills_report.pdf (accessed 17 October 2014).

Eurostat, 2014. *European Statistics.* Online, available at: http://ec.europa.eu/eurostat.

IILS, 2009. 'World of Work Report 2009: The Global Job Crisis and Beyond'. ILO, IILS. Online, available at: http://ilo.org/global/resources/WCMS_118384/lang--en/index.htm (accessed 25 November 2014).

ILO, 2013. 'The Double Dividend and Environmental Tax Reforms in Europe'. EC-IILS Joint Discussion Paper Series No. 13, ILO, EU and IILS. Online, available at: www.ilo.org/wcmsp5/groups/public/---.../wcms_194183.pdf (accessed 1 April 2015).

ILO, 2014. *Developing with Jobs.* Geneva. Online, available at: http://ilo.org/wcmsp5/groups/public/--dgreports/--dcomm/documents/publication/wcms_243961.pdf (accessed 26 July 2014).

OECD, 2010. *OECD Environmental Performance Reviews: Japan.* Online, available at: www.oecd.org/env/country-reviews/46412900.pdf (accessed 4 November 2014)

OECD, 2011. *Towards Green Growth. Summary for Policy Makers.* Paris. Online, available at: www.oecd.org/greengrowth/48012345.pdf (accessed 3 November 2014).

OECD, 2013. *OECD Environmental Performance Reviews: Mexico 2013.* Online, available at: www.oecd.org/env/country-reviews/EPR%20Highlights%20MEXICO%202013%20colour%20figures.pdf (accessed 4 November 2014).

Piketty, T., 2014. *Capital in the 21st Century.* Cambridge, MA: Harvard University Press.

Saad-Filho, A., 2010. 'Growth, Poverty and Inequality: from Washington Consensus to Inclusive Growth'. *UNDESA Working Paper* No. 100. November. Online, available at: www.un.org/esa/desa/papers/2010/wp100_2010.pdf (accessed 4 December 2014).

Slingenberg, A., Rademaekers, K., Sincer, E. and van der Aa, R., 2008. *Environment and Labour Force Skills.* Rotterdam: Ecorys. Online, available at: http://ec.europa.eu/environment/enveco/industry_employment/pdf/labor_force.pdf (accessed 2 October 2014).

WEF, 2014. *Global Risks 2014.* Online, available at: www3.weforum.org/docs/WEF_GlobalRisks_Report_2014.pdf (accessed 25 November 2014).

Wo, L., 2014. *Green Jobs in China: Comparative Analysis, Potentials and Prospects.* Friedrich Ebert Stiftung. Online, available at: www.fes-asia.org/media/publication/2012_Green_Jobs_Study_CHINA.pdf (accessed 20 September 2014).

World Bank, 2008. *Turkey Economic Memorandum: Sustaining High Growth: Selected Issues.* World Bank Report No. 39194, 10 April 2008.

World Bank, 2012. *Inclusive Green Growth: the Pathway to Sustainable Development.* Washington, DC.

World Bank, 2013. *Turkey Green Growth Policy Paper: Towards a Greener Economy.* Washington, DC. Online, available at http://documents.worldbank.org/curated/en/2013/04/17782290/turkey-green-growth-policy-paper-towards-greener-economy (accessed 10 August 2014).

World Resources Institute, 2011. *A Compilation of Green Economy Policies, Programs, and Initiatives from Around the World.* The Green Economy in Practice: Prepared for Interactive Workshop 11 February 2011. Online, available at: http://pdf.wri.org/green_economy_compilation_2011-02.pdf (accessed 4 October 2014).

6 Green jobs – stepping stones or stumbling block for sustainable development?

The concepts of green jobs, green economy and green growth have emerged from different sources, through the work of different organisations and with different target audiences, but they are used almost interchangeably. A main driver behind the development of these concepts has been the move towards a more integrated and holistic approach to incorporating the environment and development in economic decision making, policy and planning.

The concept of sustainable development is also holistic and includes economic, social and environmental considerations.

Sustainable development is development that meets the needs of the present generations without compromising the ability of future generations to meet their own needs. Green jobs are decent, environmentally friendly and low carbon. Green jobs by definition meet the needs of workers and do not compromise the ability of future generations to meet their needs. Thus, green jobs enable sustainable development.

However, the share of green jobs in total jobs is small, indicating a limited potential for influencing the economy or society as a whole. The size of the green economy roughly corresponds to the rate of decarbonisation (1.2 per cent, see Chapter 2, Table 2.1), and also to the revenue from environmental taxation.

While this correlation does not mean causality, it seems that increasing the share of the green economy could help in decarbonisation. Thus, in order to enable sustainability, the share of green jobs has to increase significantly.

The indicators allowing the identification of green jobs are extremely difficult to identify, as concepts of decent, environmentally friendly and low carbon are evolving and are country and sector specific. Technological and economic changes influence the green content: development changes the perception of decent and low carbon jobs. Still, the methodology for defining green jobs should be described in order to provide some definitional consistency. An illustrative set of indicators helps to define green jobs, relating to sector, product or service, production method, position in value chain, etc. (see Box 6.1).

The illustrative list of indicators shows that the concept of green jobs includes social, economic and environmental considerations.

Box 6.1 Illustrative list of indicators helping to define green jobs

1 The sector that a job is found within and whether it is considered to be a 'green sector', such as renewable energy and recycling;
2 The product or service's impact or output on the environment provided by the company in which the job is found;
3 The production method, including techniques and practices, used by the company to provide its product or service;
4 Green awareness and commitment to the environmental cause displayed by the company;
5 The position in the value chain of the product or service;
6 The nature of the occupational profile and whether it contributes to improving the environment in some way;
7 The specialised green skills and abilities required for the job;
8 The level of job decency and whether it provides adequate wages, job security, safe working conditions, etc.;
9 The green workload of the job that is dedicated to performing green tasks versus work in traditional areas.

Source: author's compilation based on ILO (2013)

A summary of key words from the full list of definitions (see Boxes 1.1, 2.1, 3.1) is provided in Table 6.1, categorised according to the three dimensions of sustainable development criteria.

Table 6.1 shows that there is significant overlap between the green concepts, including language around growth and economic development, environmental protection, low carbon development, resilience, resource efficiency, ecological sustainability, human well-being, inclusiveness and equity. In this regard, green jobs, green economy and green growth can be seen as tools for achieving sustainable development.

Criticisms to the green concepts relates mainly to vagueness and competition with sustainable development and internationally agreed principles. On the other hand, streamlining may attempt to treat countries in the same manner (i.e. be based on a one-size-fits-all approach), which may result in trade protectionism.

In the context of Rio+20, the relationship of green economy as a means to achieving broader sustainable development and poverty eradication was given a clear emphasis. Improving resource-efficiency, decarbonising the economy, using less (and cleaner) energy and protecting the environment are necessary to achieve sustainability.

The empirical findings show that green jobs require just slightly more skills and education than traditional jobs, and that they make between 1–8 per cent of total jobs (see Table 6.2), and have a slightly greater share in GDP.[1] The value of green sectors grows at higher rates than jobs in developed countries, while in developing countries green sectors are labour intensive and contribute more to employment than growth.[2]

Table 6.1 A summary of key words categorised according to the three dimensions of sustainable development criteria

Dimension	*Green job*	*Green economy*	*Green growth*
Social	Decent; well paid; socially inclusive; better quality of life; provide a family-supporting wage; path out of poverty	Human well-being; social equity; socially inclusive; reduced inequalities; better quality of life; social development; equitable access; addressing needs of women and youth.	Well-being; socially inclusive; access to basic commodities for the impoverished; meeting demands for food production transport, construction, housing and energy
Economic	Producing environmental goods and services	Growth in income and employment; public and private investments; resilient economy; economic growth; new economic activity	Economic growth and development; technology and innovation; environmentally sustainable economic progress; more resilient; sustained economic growth; driver for economic growth; new growth engines; green technology; new job opportunities; qualitative growth rather than simply increasing GDP; job creation or GDP growth
Environmental	Pollution management; cleaner production and technologies; resource management; contributing directly to protecting or enhancing air, water and planet	Reducing environmental risks and ecological scarcities; low carbon; resource efficient; reduce carbon emissions and pollution; enhance energy and resource efficiency; prevent loss of biodiversity and ecosystem services; within ecological limits of the planet; environmental responsibility; finite carrying capacity	Protection and maintenance of natural assets and environmental services; provision of resources and services; low carbon; using fewer resources and generating fewer emissions; resource efficient; cleaner; climatic and environmental sustainability; energy and resource efficient; minimises pollution and environmental impacts; resilient to hazards; harmony between the economy and the environment; environmental protection; reduce GHGs

Source: authors and Allen and Clouth, 2012: 61.

Table 6.2 Share of green jobs in employment

	Per cent in total employment
Mexico	4.5
Mauritius	6
South Korea	4
Spain	3
Bangladesh	1–5
Japan	1.2

Source: authors' compilation.

As regards the content, one of the main features of the green economy is that it is low carbon and clean. However, there are approaches that consider that all activities should be eventually green, provided that they are cleaner/greener than the benchmark. A narrow approach considers that green economy is part of overall economy and deals only with provision of environmental goods and services. This leads to different understandings of green growth.

A narrow perspective sees green growth as growth in the environmental industry and services – both in absolute terms and as a share of total economy. A wider approach considers greening the whole economy (i.e. making it less environmentally harmful) and the related economic growth as green growth.

Industry (supply) and consumers (demand) have a lead role in ultimately getting the economy to greener growth, the government has an equally important role in terms of creating an enabling culture that will encourage better environmental performance and the widespread adoption of practices that can lead to greener growth. The drivers of which depend on the sector and country (see Table 6.3).

In this context the biggest deficiency in approaching green management models is insufficient progress within the business sector. There are still inhumane working conditions, unfair treatment of labour, technologies that cause

Table 6.3 Drivers of green growth

EU approach	*World Bank approach*
1 An industrial policy for the globalization era	1 Pricing and fiscal policies: taxes, subsidies, or subsidy removal.
2 A digital agenda for Europe	2 Institutions, norms and regulations, and behaviour-based policies.
3 Innovation union	3 Innovation and industrial policies
4 Youth on the move	4 Education and labour markets policies
5 An agenda for new skills and jobs	5 Natural capital, agriculture, and ecosystem services management
6 European platform against poverty	6 Infrastructure, building, urbanism, transport, and energy
7 Resource-efficient Europe	

Source: authors' compilation.

high levels of pollution and overall have harmful effects on the quality of life. Therefore, the involvement of the business sector is necessary for the achievement of SDGs.

For the eradication of poverty by 2030, which is set as global development vision, SMEs are expected to have an indispensable role. They will be crucial for the employment of an additional 470 million workers that are expected to enter the labour market by then.

Big companies, especially multinational corporations, have financial strength, specific expertise and infrastructural capacities that enable expansion into other markets. In connection with small businesses they significantly contribute to employment. Some of them have integrated the issues of sustainable development into their business strategies and go beyond the traditional concept of corporate social responsibility. This is primarily accomplished through three activities:

- investments in innovation: financial capacities enable investments in innovation that enable expansion into new markets and meeting the needs of a wide range of people, including the poorest;
- promoting sustainable business practices and achieving cost competitiveness through the simultaneous preservation of soil, water, wildlife and other natural resources;
- attracting and hiring the most competent and highest quality workers and promoting workers' rights.

For fostering the role of the multinational companies in achieving goals of new global policies, their governance mechanisms should improve and enable reporting on the impact of their activities on the environment and society. This is necessary for the establishment of trust and quality partnerships with both the governmental and non-governmental sector (UN, Report of the High-Level Panel of Eminent Persons on the Post-2015 Development Agenda, 2013: 11).

The value of annual sales of the 200 world's largest corporations is 18 times greater than the annual income of the 1.2 billion people that are affected by the highest rate of poverty.[3] These corporations employ 0.78 per cent of the global workforce, while the value of their sales is equivalent to the value of 27.5 per cent of all global economic activity (Anderson, Cavanagh, 2000: 1). Thus, they have significant capacity and potential to combat global poverty. However, some of them are far from being examples of best practices.[4] On the other hand, there are companies that use a green agenda as a marketing tool, enabling access to new markets and new consumers, and present their activities as 'green' or 'sustainable', 'clean', 'eco-something', and other types of 'greenwashing'.

The national political and institutional framework has significant impact on (lack of) efforts to foster the transformation of business towards more sustainable economic models, e.g. in 2011 the governments of Denmark, China, Kenya, Korea, Mexico, Qatar and Ethiopia launched the Global Green Growth Forum (3GF) to realise the potential for long-term global, inclusive green growth. The main objective of this initiative is to establish effective cooperation between governments, business sectors, investors and international organisations to contribute to the joint efforts of achieving an inclusive green growth. The partners of the 3GF are some of the largest multinational companies (Samsung, Siemens, Danfoss, Hyundai, etc.) and international organisations (such as the OECD, the International Energy Agency and others).

Significant investment is necessary to promote green economy. The investment in eco-industries and environmental industries supports also construction, capital goods industries and associated services.[5] Compared to the rest of the economy, the rate of growth of eco-industries was much higher during the financial and economic crisis of 2008 and 2009, but slightly lower since then. Eco-industries served as an economic stabilizer during the recession. Their resistance to the crisis was among the reasons to launching green stimulus packages in developed countries. Such packages provided an impetus for greening existing industry and job creation. However, some examples indicate that such green jobs actually disappear once the subsidy programme ends. In the same period, developing countries faced a slowdown of their exports to advanced economies. They have been confronted with the need to find new sources of economic growth and job creation. Thus, the drivers of green growth are different in developing and developed countries.

The experience of developing countries shows that good quality jobs matter for development. Thus the green jobs, by definition (because they are decent and clean) favour development.

The two main challenges regarding green jobs are identified: The first is how to create the right conditions for more new jobs. Second is the fear that strict environmental obligations could jeopardise jobs and competitiveness.

Green policies are considered as those that enable the creation of a double dividend. Such policies should use regulation and market based instruments to optimise measures and achieve goals. They have evolving aspects, regarding their environmental impact (in absolute terms or compared to the benchmark), social impact (regarding decent jobs they create) and economic impact (regarding the price of products/services and competitiveness).

Green economic policies have a role in poverty eradication and achieving global sustainable development.

In emerging economies (China, Brazil) green investments were found to accelerate economic growth and job creation. Compared to business-as-usual levels, job creation is even faster than growth.

A growth model leading to the establishment of the green economy is seen as the way to stimulate job creation and to improve the quality of existing jobs. Each country should consider the implementation of green economy policies in a context of sustainable development and poverty eradication. Such policies should lead to sustainable growth and job creation, particularly for youth, women and the poor.

Greening the economy mainly relies on domestic jobs, and to a certain extent, foreign markets. Strategies for green urban development are basically local development strategies that rely on local jobs. Thus, the transformation to a green economy can be seen as anti-globalisation. The initiatives lobbying for greening the economy are, at least to certain extent, anti-global. They are designed and implemented in various frameworks – within green growth strategy, low carbon and/or sustainable development strategies.

The evidence suggests that gains and losses in terms of jobs, inclusion and equality are not happening by default, but neither can they be programmed with one-size-fits-all policy recipes. The shaping of the right mix of policies is very much country-specific. Countries and regions with large shares of resource-intensive and high-emitting industries face different challenges from those with a lighter legacy of unsustainable production patterns. The latter may have labour markets dominated by sectors exposed to climate change such as agriculture.

The overall conclusion is that the 'green' growth is sustainable – i.e. it supports economic growth, social inclusion and well-being, provides enabling conditions for employment and fair labour and protects the natural resources and ecosystem. However, there is no consensus on how to drive the growth. Designing the strategies for green growth is thus even more challenging. Thus, the development of green strategy can be considered as an indicator and driver of development, both.

Its impact depends on the time horizon. The further away the horizon, the more adjustment can be done.

From a different analytical perspective, green jobs are a result of green policies/green entrepreneurship. According to such an approach, the employment impacts of green policies relative to a baseline are examined, e.g. what the employment impacts of the EU 20–20–20 Strategy are. These issues are crucial for estimating the overall labour market impact of green policies, including the impact of fiscal stimulus packages. Some estimates include only directly created jobs (e.g. production of equipment, maintenance), while other include indirect (in supply chain) and induced jobs (created by the consumption of the directly and indirectly employed).

Most of the studies estimate that the number of green jobs is expected to grow. However, there is limited evidence on the evolution of job quality. Greening the economy per se will not improve the quality of indirect or induced jobs – the changes in job quality depend on the types of jobs created and those lost.

The green economy is in a position to grow quickly, but the relatively small size of the green economy suggests that a majority of the jobs that will be created during recovery are likely to come from the production of products and services outside of the green economy.

While green jobs are seen as a means to create more and better jobs and to contribute to sustainable development, they are often considered as expensive to create. Also, if the creation of a green jobs leads to a loss in traditional jobs, such development might lead to rising differences. On the other side, the recovery of the world's largest economies can be facilitated by transitioning to a 'green economy'.

Green growth should help in dealing with the systemic issues of jobless growth, inequality and income disparity. The developing countries need to create more than 50 million new jobs a year. A new generation of green jobs, or sustainable livelihoods, could play a role in this, although they won't solve the problem alone.

A number of studies have estimated the number of green jobs and growth potential. A range of methods, reflecting the different definitions of green jobs were used. These estimates are not comparable due to:

- different geographical coverage (United States, EU member states, global);
- sectoral differences (often focusing solely on individual sectors or subsectors);
- gross and net effects and the inclusion/exclusion of indirect and induced jobs; and
- different assumption concerning economic growth and the effect of existing business-as-usual policies.

The heterogeneity of individual economies makes drawing broad conclusions very difficult. Findings suggest that climate change policies in general and renewable energy in particular can generate considerable employment. On the other hand, there is the potential for job destruction in non-green industries. Green jobs are seen as employment opportunities and tools from one longer-term growth path to another. This is particularly important in times of crisis, when involuntary unemployment is widespread. Green jobs are drivers of growth and can be used as indicators of development.

Given the current lack of an adequate quantitative modelling paradigm for environmental policy analysis, the effectiveness of such policy interventions and their economic impacts as well as the trade-offs they involve are not well-known. There is a strong need therefore for the design and utilisation of analytical models for environmental policy analysis, which would allow more accurate predictions of the private benefits of investments and consequently allow public policy to be calibrated to leverage private sector reactions through a combination of price and other incentives.

The strategic policy frameworks that include sustainable development issues (e.g. generic NSDSs, sustainable development included in other policy strategies, etc.) vary significantly among the countries and world regions and are largely dependent on the respective political-administrative systems, economic development, environmental conditions, institutional capacities, political culture, etc.

Generally, the UN is an important driver for the development/implementation of NSDSs or addressing sustainable development more strategically. The UN developed guidelines for NSDSs and also acts as a support organisation, particularly for lesser developed countries. For instance, UNEP is supporting the development and implementation of NSDSs in the Asia-Pacific region and was involved in the establishment of the ILAC.

The external support of the UN and donor organisations (e.g. the Norwegian Agency for Development Cooperation in the Asia-Pacific region) has many advantages. However, experience shows that this external support is currently accompanied by weak national ownership and challenges in establishing effective governance systems in the respective countries.

'Generic NSDSs' can only be found in a limited number of countries in the world regions covered (e.g. Thailand, several Pacific Island countries); several other countries have adopted national Agenda 21 strategies (e.g. China, Philippines, Brazil). Nevertheless, sustainable development is increasingly included in strategic policy documents, most often in NDPs (e.g. India, Mexico, Costa Rica). Moreover, environmental policy plans are still dominant in several countries, some of them also address cross-sectoral issues (e.g. Japan, Egypt).

Despite efforts to address economic, social and environmental issues in sustainable development strategies, environmental sustainability is in many countries the prevailing approach.

Several NSDSs or strategic policy documents that include sustainable development issues refer to cultural aspects or national heritage and governance mechanisms as important issues.

Some regional commitments to develop NSDSs or integrated sustainable development approaches have emerged in several world regions, e.g. the Pacific Island Forum, the Latin America and Caribbean Initiative for Sustainable Development and the MSSD.

Although sustainable development is increasingly taken up by various countries in the different world regions (UN ESCAP, 2008; Meadowcroft, 2007), major challenges remain, e.g. concrete actions on the ground, integrating sustainable development into all policy sectors, coordination of sustainable development objectives among government levels as well as developing comprehensive and effective monitoring and evaluation processes.

Investing in renewable energy, improving efficiency and altering technologies are important for achieving a green economy. The moves towards a green economy are considered as a strategic economic policy agenda for achieving sustainable development. A green economy recognises that the goal of sustainable development is improving the quality of human life within the constraints of the environment, which include combating global climate change, energy insecurity and ecological scarcity.

For implementation, specific enabling conditions are required. These consist of national regulations, policies, subsidies and incentives, as well as international market and legal infrastructure, trade and development aid. This is being done through general and specific rules.

At the EU level, enabling conditions for greening the economy include market instruments and regulation (EU Emission Trading System, Effort Sharing Decision).

Despite the development of technologies and a legal framework that supports greening the economy, there is a strong empirical anomaly, which shows that highly profitable energy-saving opportunities exist, yet the technologies embodying these opportunities have not spread universally through the global economy.

Notes

1 The percentages indicating the penetration of environmental jobs can be used as a reference, but they cannot be compared between countries due to the diversity of methodologies used to calculate green jobs.

2 This is a preliminary finding and should be elaborated on in further research.

3 According to the World Bank definition, these are individuals who survive on less than US$1/day.

4 For instance, Exxon, the world's third largest company in terms of revenue and the sixth company according to the amount and toxicity of pollutants emitted into the air in the United States of America (Political Economy Research Institute of the University of Massachusetts, 2013) allocates less than 1 per cent of the realised profit for research on the use of renewable energy and environmental protection (Weiss, Kougentakis, 2009). Shell is found responsible for direct environmental pollution in Nigeria, Argentina and Brazil, and for human rights violations (Amnesty International, 2012).

5 The level of investment in the EU in the 1990s was €54 billion each year (Ecotec, 2002).

References

Allen, C. and Clouth, S., 2012. 'A Guidebook to the Green Economy'. Issue 1: *Green Economy, Green Growth, and Low-carbon Development – History, Definitions and a Guide to Recent Publications.* Division for Sustainable Development, New York: UNDESA.

Amnesty International, 2012. *Amnesty International Report 2012: The State of the World's Human Rights.* London: Amnesty International.

Anderson, S. and Cavanagh, J., 2000. *Top 200, the Rise of Corporate Global Power.* Washington, DC: Institute for Policy Studies.

Ecorys, Teknologisk Institut, Cambridge Econometrics, CESifo, Idea Consult, 2009. 'Study on the Competitiveness of the EU Eco-industry'. Brussels. Online, available at: http://ec.europa.eu/environment/enveco/eco_industry/pdf/report%20_2009_competitiveness_part1.pdf and http://ec.europa.eu/environment/enveco/eco_industry/pdf/report%20_2009_competitiveness_part2.pdf (accessed 2 October 2014).

Ecotec, 2002. 'Analysis of the EU Eco-industries, their Employment and Export Potential'. Birmingham. Online, available at: http://ec.europa.eu/environment/enveco/eco_industry/pdf/main_report.pdf (accessed 27 November 2014).

European Commission, 2011. 'EU Environment Policy Supporting Jobs and Growth'. Luxembourg: Online, available at: http://ec.europa.eu/environment/enveco/industry_employment/pdf/facts_and_figures.pdf (accessed 16 November 2014).

GHK, Cambridge Economics, and Institute for European Environmental Policy, 2007. 'Links between the Environment, Economy and Jobs'. London. Online, available at: http://ec.europa.eu/environment/enveco/industry_employment/pdf/ghk_study_wider_links_report.pdf (accessed 23 November 2014).

GHK, 2009. 'The Economic Benefits of Environmental Policy Global Climate Network, Low Carbon Jobs in an Interconnected World'. Online, available at: www.ippr.org/publications/low-carbon-jobs-in-an-interconnected-world (accessed 2 December 2014).

ILO, 2013. 'Sustainable Development, Decent Work and Green Jobs'. Report V. Online, available at: www.ilo.org/wcmsp5/groups/public/---ed_norm/---relconf/documents/meetingdocument/wcms_207370.pdf (accessed 3 October 2014).

ILS, 2009. *World of Work: the Global Job Crisis and Beyond.* Geneva: ILO.

ILS and Chinese Ministry of Human Resources and Social Security, 2010. Study on Green Employment in China. Beijing.

Meadowcroft, J., 2007. 'National Sustainable Development Strategies: Features, Challenges and Reflexivity'. European Environment Special Issue: Sustainable Development Strategies in Europe: Taking Stock 20 years after the Brundtland Report, 17 (3): 152–163.

Millennium Institute, 2012. 'Growing Green and Decent Jobs'. ITUC. Online, available at: www.ituc-csi.org/IMG/pdf/ituc_green_jobs_summary_en_final.pdf (accessed 19 November 2014).

UN, 2012. 'A Guidebook to Green Economy'. Online, available at: http://sustainabledevelopment.un.org/content/documents/GE%20Guidebook.pdf (accessed 5 November 2014).

UN, 2013. 'Report of the High-Level Panel of Eminent Persons on the Post-2015 Development Agenda: A New Global Partnership: Eradicate Poverty and Transform Economies through Sustainable Development'. Online, available at: www.un.org/sg/management/pdf/HLP_P2015_Report.pdf (accessed 10 January 2015).

UN Environment Management Group, 2011. 'Working Towards a Balanced and Inclusive Green Economy' (4). Online, available at: www.unep.ch/etb/pdf/2009%20statement%20deliver%20as%20one/Interagency%20Joint%20Statement.%20E%20rev1.pdf.

UNEP, UNCTAD, UNDESA, 2011. 'Transition to a Green Economy: Benefits, Challenges and Risks from a Sustainable Development Perspective'. UN

UNEP, UNCTAD, UN–OHRLLS, 2011. 'Why a Green Economy Matters for the Least Developed Countries'. UNCSD Secretariat. Papers for the Rio+20 Conference including A/CONF.216/PC/6 and A/CONF.216/PC/1.

UN ESCAP, 2008. 'Greening Growth in Asia and the Pacific'. Bangkok: UN. Online, available at: https://sustainabledevelopment.un.org/index.php?page=view&type=400&nr=783&menu=1301 (accessed 15 January 2015).

Weiss, D.J. and Kougentakis, A., 2009. 'Big Oil Misers'. Online, available at: www.Americanprogress.org/issues/2009/03/big_oil_misers (accessed 21 December 2014).

7 Conclusions

The key challenge of this book is based on the adoption of concepts – baseline frameworks and green concepts.

The main distinction between a 'classical' and a 'green' approach relates to quality of living and working conditions and respect for the environment. While these are generally accepted goals, there are still inhumane working conditions, unfair treatment of labour, working poverty, technologies that cause high levels of pollution and overall have harmful effects on the quality of life. Therefore, the involvement of the business sector in greening the economy has a strong potential to improve the implementation of green concepts.

The conceptual framework of green terms (job, economy, growth, policy) is not uniform. A number of theses and considerations on what is actually green exist.

Differences in the width of the concept are key: for example, a generally accepted idea is that green economy relates to activities in the energy sector related to renewable energy and climate change, while wider green concepts include green jobs (which might be outside these sectors), green growth and green policies. Even if the global expert community agrees on definitions, the question is whether through the realisation of proposed concepts the distinction of key terms, such as growth, jobs and the economy is achieved. Table 7.1 summarises the interpretation of key terms – within both the classical and green approach.

In recent literature, green concepts are mentioned as the basis for sustainable development in the future. In addition, there is the concept of green development as a logical step for greening the global future through the implementation of green growth based on green jobs. However, green development is mainly linked to the concept of land use, rather than the overall growth, which should be inclusive and environmentally sound with the aim to reduce poverty and build a shared perspective (see Table 7.2).

Regardless of the terminology and conceptual differences, the analysis of the implementation and operationalisation of some green concepts through a variety of national economies still mainly point to the fact that the creation of green jobs is necessary, but not sufficient, to enable sustainable development. The case studies provided clearly show that although the green concept is well-intentioned, its

Table 7.1 Summary of key terms – classical and green approach

Term	*Classical approach*	*Green approach*
Job	A set of work tasks performed by people in order to gain resources for living.	Any job that produces products or services related to environmental conservation and management, makes work processes more sustainable and has decent work conditions (ILO, 2010).
Economy	Human activity consisting of three main factors: production, consumption and trade.	Resilient economy that provides better quality of life for all within ecological limits of the planet (Green Economy Coalition, 2011).
Growth (economic)	The increase in the value of goods and services in the economy (increasing wealth of the nation).	New model of economic growth (targeting poverty reduction, job creation and social inclusion) and environmental sustainability (climate change mitigation, biodiversity loss and access to drinking water and clean energy) (Global Green Growth Institute, 2012).
Policy (public)	Activity of a government in response to social problems as a condition to be changed. All that a government chooses to do, or not to do.	Measures for integrating environment and development through effective legal and regulatory frameworks, economic instruments, market and other incentives and systems for integrated economic and environmental accountings.

Source: authors, based on Wikipedia for classical definition and the selection of terms offered in this book for green approach.

Table 7.2 Classical, sustainable and green development concepts

Term	*Classical approach*	*Sustainable approach*	*Green concept*
Development	The process of improving the living standards and welfare of the population.	Development that meets the needs of the present without compromising the ability of future generations to meet their own needs.	Land use planning concept that includes consideration of community wide or regional environmental implications of development, as well as site-specific green building concepts.

Source: authors.

implementation depends on local circumstances – economic, political and social. Developed countries perceive green growth as a way to create new markets and demand, while developing countries rely more on labour intensive growth and less expensive green jobs. It seems that economic considerations have precedence over environmental considerations in all countries analysed.

The differences in approaches limit the possibilities for quantitative analyses and cross-country comparison. Still, a qualitative assessment is possible.

In general, Europeans believe that green jobs lead to a green economy that promotes green growth and creates a framework for sustainable development. The operationalisation of the theoretical framework underlying green concepts is implemented through the measures of green policies. Conversely, Asian countries use green growth as a starting point for creating green jobs and achieving sustainability.

Also, it can be identified that the drivers of greening an economy are different in developed and developing countries. In developing countries, greening is mostly motivated by scarcity or resources and environmental degradation (issues related to natural capital), while technological development and innovation drive the process in developed countries (see Chapter 6, Table 6.3).

The common features relate to the necessity to change the patterns of production and consumption.

Green jobs are considered to be the backbone of a greener sustainable future, based on the green economy and green growth. Although analysts often associate them with activities in low carbon industries, today they are becoming more and more considered as green jobs in the service sector. Out of these jobs, regardless of whether they are light green or dark green jobs as in Denmark, or green-collar jobs as in America, the creation of new jobs in the future and the improvement of social well-being is expected. Moreover, it is expected that green jobs will become the drivers of national economies and the drivers of economic growth.

Whether such structural changes in the labour market are possible and whether the expectations of a win–win situation for greener growth are justified are the questions that still do not have a significant desired positive response. It is hard to expect that brown industries will turn into green and sustainable industries without significant long-term financial investments in the infrastructure, the capacity building of institutions, public opinion and new knowledge and skills for green jobs.

The fact is that employment in green jobs is continuously growing. This trend is mainly the result of more rigorous environmental legislation in most countries in the world and constant new international initiatives, environmental contamination, remediation, management and monitoring in the sectors of waste, water, air and nature protection. In the environmental and development policies of the countries analysed in this book, the number of green jobs is increasing. Particularly, it relates to jobs in the renewable energy sector, but also to other sustainable activities. The area for further growth of green jobs is, in most countries, linked to the development of sustainable tourism, in particular in nature protected areas, forests, coastal zones and the like. Most green jobs are mainly characterised by better compensation packages, safer and more decent jobs. Furthermore, most people who have green jobs have an advanced level of skills, thereby justifying the expectations of future development drivers and bearers of green growth.

Green economies are becoming a development platform from the emergence of the economic crisis in 2008, as part of the economic recovery efforts. Although there is also no generally accepted definition for the concept of a green economy, all explanations include the tendency for simultaneous realisation of human well-being and the investigation of environmental risks.

In the last few years, green economy has been recognised worldwide as a vehicle for sustainable development, especially following the Rio+20 Conference, where green economy within the context of sustainable development and poverty eradication was one of the two main themes. An inclusive green economy that is adapted to national circumstances has the potential to advance poverty eradication and improve human well-being, while also limiting, repairing and even reversing damage to ecosystems. It can lead to more intelligent product design, greater resource efficiency, reductions of GHG emissions, less waste and pollution, and greater economic opportunities for poor and vulnerable groups. The opportunities of an inclusive green economy vary by country.

Among the European Union countries there is also no single approach to a green economy. For example, in the United Kingdom, policies go towards the direction that the whole economy needs to be green. Furthermore, Denmark has branded its businesses and living standards as green – perhaps most of all the countries of the world. However, in some other EU countries the economic progress and the raising of living standards of the population are still considered as more important than the greening of economies (e.g. Bulgaria, Romania).

Unlike in the EU, green economy in the United States is mainly associated with greening the energy sector. It is similar in Japan, where the energy sector and the air pollution control sector are the fastest growing sectors, which make the bulk of green economy of the country.

In developing countries the green economy sectors have risen by over 20 per cent in the last few years. Considerable development capacities have been recognised in sustainable activities in the sectors with significant greening potential (agriculture, forestry, tourism, renewable energy and low carbon industries).

In the transition to a green economy, developing countries face financial, technological and capacity challenges. They require support from the international community in terms of capacity building, technology transfer, funding and technical support. Furthermore, investments in green economic sectors need to be productive, creating decent jobs that offer paths to skill and knowledge. Green economy concepts are associated with low carbon and energy efficiency activities and also the innovation, R&D, and a higher level of management skills.

Investment in the green economy can foster green growth. As with other green concepts, there is no consensus on the contextual framework of green growth. It can be described as an economic development and growth instrument for the efficient use of natural resources, the prevention of environmental degradation, increased social welfare and employment, the reduction of poverty and the promotion of innovative, efficient and clean technologies. The green growth focuses on attracting investment, the green economy targets wider and deeper reform to create an economic system that better serves society. Thus, the green growth is not necessarily the way to achieve the green economy, whose objectives are to meet various needs – for food, transport, energy and so on – in a sustainable and equitable way.

While there is some significant overlapping (Table 7.1) the definitions of green jobs, green growth and green economy are not coherent. While green jobs are (according to international labour statisticians) environmental jobs that are decent, a green economy can include all sectors (not only environmental ones). Thus, some of the workers in a green economy do not necessarily have green jobs, but do they perform green work (see Chapter 1, Table 1.14). Thus, the number of green jobs and their value added does not necessarily correspond to the size of the green economy, which illustrates one of the major shortcomings of the 'green' concepts – they are not comparable.

The fact is that people tend to like the pleasant life and prosperity brought about by economic growth, regardless of the limitations posed by the environment. Therefore, the harmonisation of the pursuit of economic growth with environmental sustainability has been an imperative for two decades, which was globally defined by various international organisations as green growth. Green growth has become the new model of economic growth, and national economies should prepare their green growth strategies and implement green growth plans as well as promote and support high-quality research activities to support these processes. Green growth should enable sustainability and gradually replace 'brown growth' that brings short-term benefits but destroys natural systems and thus diminishes prosperity.

Unfortunately, because of the cachet that consumers give to environmentalism, some industries have sought to present their dirty and polluting activities as 'green' or 'sustainable', with marketing ploys like the use of words like 'clean', 'green' or 'eco-something', and other types of 'greenwashing'.

Green growth should also provide for reducing the gap between developed and developing countries. The operationalisation of such an ambitious goal is difficult to achieve. Decarbonisation processes that are the essence of the environmental component of green growth are not always the key development objective of national economies. Therefore, conceptual directions of future green growth go in the direction of strengthening the green services and green consumption in parallel with processes of reduced saving and investment. This growth can be a stimulus for new jobs in sectors that have not been identified as possible drivers of development, such as tourism, culture and traditional sectors in the new 'green' attire, such as sustainable agriculture, sustainable forestry, manufacturing, small crafts and entrepreneurship.

China and Turkey, which had in recent years obtained the highest rates of economic growth, are among the most important pollutants. They are traditionally the most important producers of the iron and steel industry, which are the largest industrial emitters of CO_2 in the world.

Incoherence of economic development and green theory can be seen in documents at the global level over the past ten years. Green actions/tasks are explicitly contained in most sustainable development strategies. Green, as a development challenge, is not part of the MDGs, but is mentioned in the new framework of several SDGs for the post-2015 period.

While both concepts, sustainable development and green growth, integrate economic, social and environmental objectives, the terms do differ.

In order to achieve sustainable development patterns it is necessary to 'green' development policies. Almost all policy actions can be green and can contribute to sustainable development (development, social, economic, etc.). It is essential that the policies allow the protection of the environment and bring economic benefit. Thus, green policies include macroeconomic and microeconomic policies alike. They should also enable the correction of market failures (by use of market based instruments for environmental externalities).

Green policies should support the transition to a green economy and deal with the costs of transition (both direct and indirect). These include green employment (and related skills) policies, growth policies (including taxation), as well as other policies with a development dimension (e.g. science and innovation). Also, some traditional sector-specific policies might have green elements (e.g. green energy, transport policy).

Moreover, for green policies it is necessary to ensure the availability of information for the preparation and implementation of the policy decision making process and to inform the public about the process. In preparing and implementing green policies one must insist on checking the cross-sectoral actions of some of the proposed measures using a multi-sectoral or input–output analysis. Only this approach allows a multilevel analysis of the possible consequences of development decisions on the well-being of society as a whole, which is ultimately the purpose of the 'greening' itself.

In conclusion – there is a series of thesis and development concepts associated with a 'green' future. In addition to a clearly defined concept of 'sustainable development' all other green concepts have dozens of similar, yet different, interpretations, which are treated differently in different countries and disable more serious comparative analysis.

Regardless of the promising development perspectives, green economy and green growth as the challenges of the future are only indicated in the global development goals. Neither the MDGs nor the post-2015 SDGs are explicitly tied to green development as a global perspective. Some of the proposed SDGs are green, but not all of them. Thus, it can be claimed that green growth is necessary but not sufficient for sustainable development.

Therefore, an understanding of this subject can contribute to the perception of detachment from 'green' as the only energy orientation, towards 'green' through all sustainable activities and green lifestyle. Such improvements will enable the introduction of green practices and green instruments in all development policies and contribute not only to the low carbon economies, but also to the progress of the society as a whole. New knowledge and skills, new investments in innovation and new technologies, as well as new economic sectors with sustainable services open up vast opportunities for new green jobs, for the development of national green economies and green growth, for more satisfied people and the planet preserved for future generations.

This can be illustrated in that European strategy on the sustainable development of cities was developed in 2008, while the green urban issues are a more recent concept.

The perfect economic growth/social model does not and will never exist. An obsession with growth targets, at the possible expense of society, could prove dangerous and harmful. In developing the growth and social development path, policy makers should ask many questions and especially the most fundamental: 'what kind of society do we want?'

The answer is probably green, but unclear.

Index

Page numbers in *italics* denote tables.